D0151489

The Cambridge Atlas of the
Middle East & North Africa

The Cambridge Atlas of the
Middle East & North Africa

Gerald Blake · John Dewdney · Jonathan Mitchell

Department of Geography, University of Durham

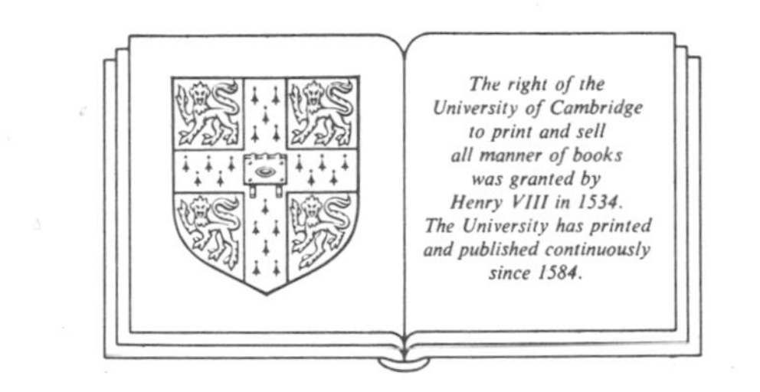

Cambridge University Press

Cambridge · New York · New Rochelle · Melbourne · Sydney

Published by the Press Syndicate of the University of Cambridge
The Pitt Building, Trumpington Street, Cambridge CB2 1RP
32 East 57th Street, New York, NY 10022, USA
10 Stamford Road, Oakleigh, Melbourne 3166, Australia

First published 1987

Printed in Great Britain by BAS Printers Ltd., Over Wallop

British Library cataloguing in publication data

Blake, Gerald H.
The Cambridge atlas of the Middle East and North Africa.
1. Middle East – Maps 2. Africa, North – Maps
I. Title II. Dewdney, John C. III. Mitchell, Jonathan
912′.56 G2205

Library of Congress cataloguing in publication data

The Cambridge atlas of the Middle East and North Africa
Bibliography.
1. Middle East. 2. Middle East – Maps.
I. Blake, Gerald Henry.
II. Dewdney, John C. III. Mitchell, Jonathan, 1959–
DS44.C36 1987 912′.56 87–6548

ISBN 0 521 24243 6

Contents

Maps

Figures

Tables

Preface

In the mid 1980s the eyes of the world focus more sharply than ever upon the Middle East and North Africa as a region of political turmoil, economic opportunity and cultural distinctiveness. Since the early 1960s great strides have been made in providing politicians, businesspeople and scholars with literature and statistical data upon which an informed understanding of the region and its peoples can be based. So far, however, there has been no comprehensive atlas available which combines accuracy and detail with explanatory comment on the geographical patterns displayed. Having worked on the production of this atlas at intervals since 1980, we now have some idea as to why such a publication has not been attempted before. Apart from problems of data, there were obvious difficulties in reconciling clarity and detail with our desire to illustrate the region as a whole. We also wanted each map to be accompanied by an informative text which would date less rapidly than the kind of information presented in yearbooks, but in a region of rapid economic, political and social change this was not always easy.

Our colleague Howard Bowen-Jones, formerly Director of the Centre for Middle Eastern and Islamic Studies at Durham University, first suggested the preparation of this atlas, and we gladly acknowledge his advice and encouragement at the planning stage. As work progressed we were fortunate to be able to draw upon the expertise of a number of colleagues in Durham University to write commentaries on our maps. These include Ewan Anderson (Map 54), John Chitham (Map 19), Michael Drury (Map 18), Carl Grundy-Warr (Map 58) and John O'Reilly (Map 51), all of the Department of Geography. Richard Lawless (Maps 14 and 15) and Ian Seccombe (Maps 24 and 25) of the Centre for Middle East and Islamic Studies; William Hale (Maps 16 and 17) of the Politics Department, Bill Williamson (Map 20) of the Department of Sociology, and Rodney Wilson (Maps 37, 39, 40, 41) of the Department of Economics also contributed, together with Alasdair Drysdale (Maps 33, 34, 35) of the Department of Geography, University of New Hampshire. We are extremely grateful to them all for their willing collaboration. Over several years of preparation we called upon the professional skills of many members of the technical and secretarial staff in the Department of Geography, and our Senior Cartographer, Arthur Corner, was most generous with technical advice and assistance. Margaret Bell supervised most of the typing with skill and patience. Outside the department, Joyce Mitchell helped draft some of the maps in their initial stages, and Heather Bleaney and Avril Shields at the Middle East Centre Documentation Unit gave unstintingly of their time and knowledge. But even with all this undeserved support and encouragement this atlas would not have been completed without the enthusiasm and gentle persistence of Robin Derricourt and his colleagues, especially Dale Tomlinson and Michael Holdsworth, at Cambridge University Press, who gave timely help and advice even when we fell short of their high expectations. We also owe a great deal to the skilful and careful editing of Elizabeth O'Beirne-Ranelagh. Lastly, thanks are also due to Colin McCarthy and his team at Thames Cartographic whose cartographic skills are evident in the following pages.

The bibliography lists as far as feasible the major sources of information and inspiration for the 58 maps in the atlas. We have also given some suggestions for further reading under each of the texts, which generally indicate the chief sources for the commentaries.

Wherever possible, the spelling of place names follows *The Times Atlas of the World*.

Gerald Blake
John Dewdney
Jonathan Mitchell

Durham, January 1987

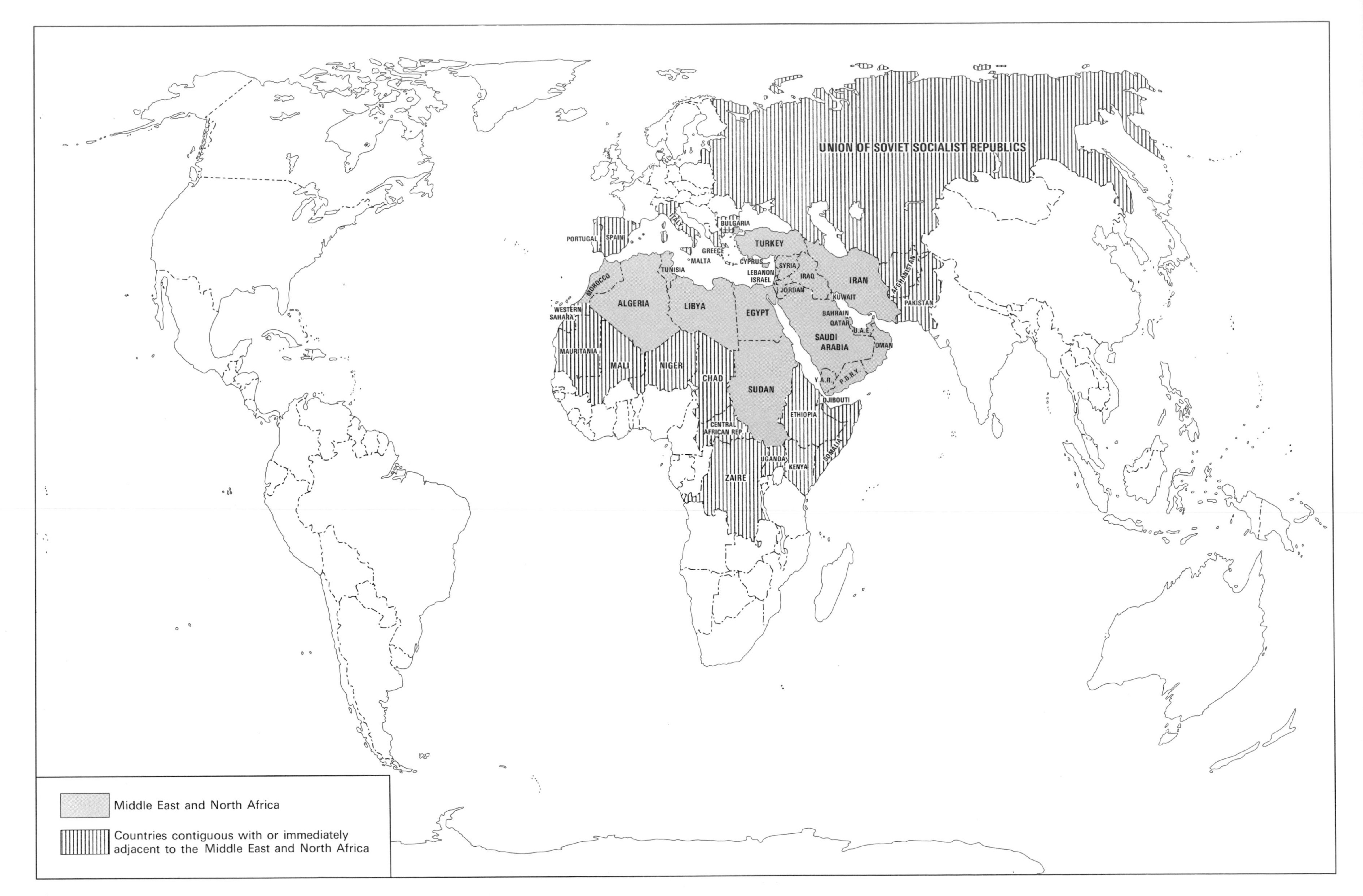

1 The Middle East and North Africa in the world

Global perspectives

Gerald Blake

There is no standard definition of either the Middle East or North Africa. The term Middle East seems to have been first used by the British in the late nineteenth century with reference to a region around the Gulf, somewhere between the 'Near East' and the 'Far East'. During and after World War II, the Middle East expanded to take in all the Arab states of Southwest Asia, with Turkey, Iran, Israel and Cyprus. Afghanistan is sometimes included, and in many definitions Sudan, Egypt, and Libya are regarded as Middle Eastern states. The term is now widely used throughout the world, although it is derived from a Eurocentric view of the map. 'North Africa' commonly comprises all the coastal states of northern Africa, plus Sudan which shares the Nile basin with Egypt. The 22 states covered in the *Cambridge Atlas of the Middle East and North Africa* are shown in Table 1. Among them are several of the larger and a number of the smallest states in the world. Together, they cover about 10.3 per cent of the earth's land surface and support 5.8% of the world population (1984).

The region is vast, extending 8370 km from east to west and 4180 km from north to south, and through 120° of longitude and 38° of latitude (Map 1). It embraces states which are markedly diverse in area, population, resources and political institutions but which can be justifiably grouped together on historic (Maps 14 and 15) or linguistic (Map 18) or religious (Map 19) grounds. They also display similar development problems, notably those associated with aridity (Maps 9–12) and dependency (Maps 40 and 41). With few exceptions they have all experienced colonial rule in the recent past. The North–South divide between the developed and developing worlds has been strikingly located along the northern fringe of the region, suggesting perhaps a new role for it at the interface between North and South (see Fig. 1). By most criteria all the states of the region, apart from Israel, belong to the Third World, though there are some anomalies such as the high level of G.N.P. per capita in oil-rich states (Map 39).

Even without substantial oil resources, the Middle East and North Africa would command considerable geopolitical importance by virtue of global location and the shape of their territory. Fringed by five seas and by two oceans, the region is easily accessible by sea. It also provides a land bridge between Africa and Asia and short sea crossings between Africa and Europe and Europe and Asia. The Middle East is located on the most convenient routes by land, sea and air linking the populous regions of Europe and North America with the populous states of the Far East (Maps 2A, 42–45). The region was seen as a vital crossroads by the Europeans long before modern transport came on the scene, and wars were fought over control of routes through the region. In the future, road links may become more important. A second Bosporus bridge is planned, and there is talk of a fixed link across the Strait of Gibraltar (Map 51). States which share a common land frontier with the region extend far into the continents of Africa and Asia (Map 1). Historically, interaction with these territories has at times been intense, involving trade, especially in North Africa, and the incursions of invading tribesmen, especially into Southwest Asia. The closeness of Europe also needs to be stressed. This geographical fact above all explains the long history of European intervention in the Middle East, and European colonisation of North Africa in the nineteenth and twentieth centuries (Maps 16 and 17).

Culturally, the Middle East and North Africa are part of the world of Islam (Map 2B). As the region which gave birth to the religion of Islam, and as the home of the holy cities of Mecca and Medina (Map 19), the Middle East has a special place in the eyes of Muslims, symbolised by the pilgrimage to Mecca (or Haj), which attracts more than a million of the faithful each year. Only about a quarter of the world's 800 million Muslims live in the Middle East and North Africa; far larger Muslim populations live in Indonesia, Bangladesh, Pakistan and India than in any state of the region. There are approximately 40 million Muslims in the Soviet Union, about one in six of the population. About half the world's official Islamic states are in the Middle East and North Africa. Cyprus, Israel and Lebanon are notable exceptions. Fewer than 7% of the population of the region are non-Muslims (Christians, Jews and others), but their social and political significance is considerable. Although Islam is a dominant unifying element, its adherents speak a variety of languages apart from Arabic, and are sharply divided between the Shi'ite and Sunni branches of the faith (Maps 18 and 19).

Politically, the region is represented in both the major power blocs, with some states professing non-alignment (Map 2C). The United States and the Soviet Union have eagerly sought client states in the Middle East and North Africa in the last 25 years, with varying fortunes. A combination of factors, including global communi-

Table 1. *State areas*

Rank		Area: Sq. km	Area: Sq. miles
1	Sudan	2,505,825	967,500
2	Saudi Arabia	2,400,930	927,000
3	Algeria	2,381,745	919,591
4	Libya	1,759,540	679,358
5	Iran	1,648,000	634,296
6	Egypt	1,000,253	386,198
7	Turkey	779,452	300,947
8	Morocco*	458,738	177,117
9	Iraq	438,446	171,267
10	P.D.R. Yemen	336,829	130,069
11	Oman	271,950	105,000
12	Yemen A.R.	195,000	73,300
13	Syria	185,680	71,772
14	Tunisia	164,150	63,362
15	Jordan†	96,000	37,065
16	United Arab Emirates	92,100	32,300
17	Israel†	20,700	7,993
18	Kuwait	17,818	6,880
19	Qatar	11,000	4,247
20	Lebanon	10,400	4,015
21	Cyprus	9,251	3,572
22	Bahrain	660	255
	(Gaza Strip)	360	139

*Excluding Western Sahara.
†Pre-1967 boundaries.
Source: John Paxton (ed.), *The Statesman's Yearbook 1984–85* (Macmillan, London, 1984).

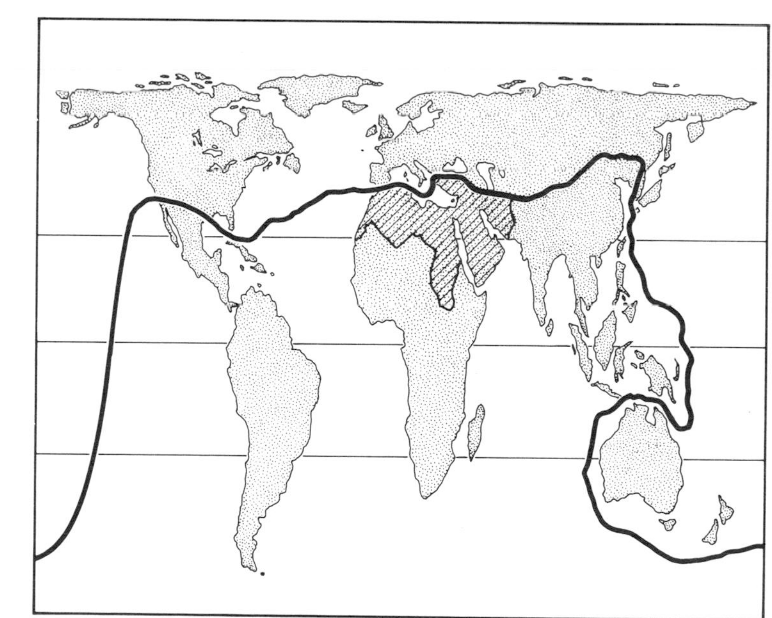

Fig. 1 The North–South divide on the Peters projection.

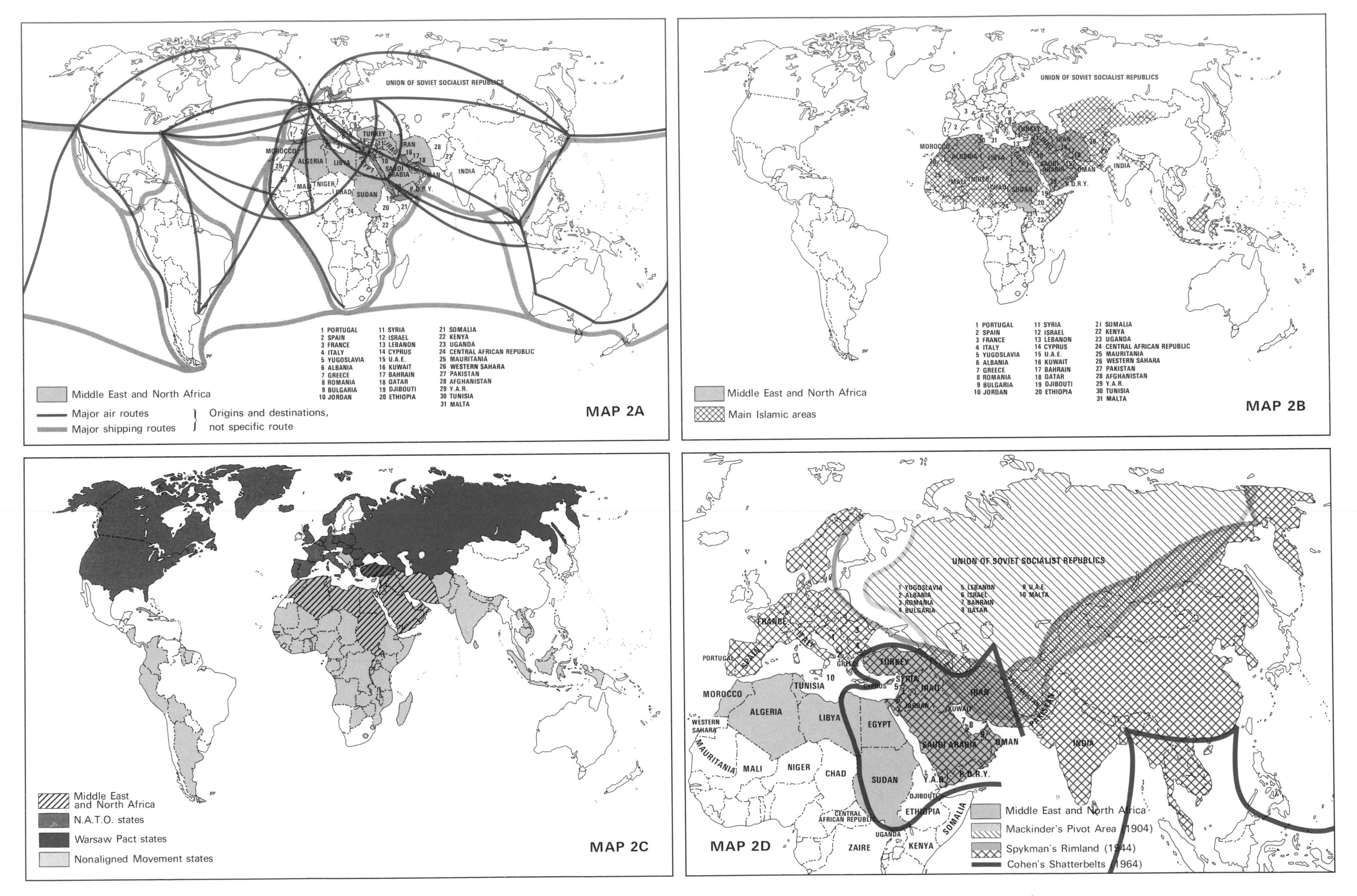

2 Global perspectives: A. external links; B. the Islamic world; C. political alignments; D. geopolitics

cations, the oil trade (Map 46), and geographical proximity to the Soviet Union and Western Europe, ensures that both superpowers currently give the region high priority. The Soviet Union is naturally interested in what happens along its southern flanks, and is much preoccupied with minimising United States influence there. The geopolitical importance of the region's strategic waterways is illustrated elsewhere (Maps 47–51), but it may be noted here that Turkey is a member of the N.A.T.O. alliance and controls the Soviet Black Sea fleet's access to the Mediterranean through the Turkish Straits (Map 49). The intensity of superpower rivalry is reflected in the high level of arms sales to client states, and the pattern of bases and alliances (Map 54). The United States has military facilities in several states in strategically favourable locations, notably Morocco, Egypt, Turkey and Oman, but a future change of regime could quickly end these arrangements.

Several geographers and historians have attempted to develop world views of geopolitical relationships, and in all of these, parts of the Middle East and North Africa figure prominently. Although their ideas are defective in certain respects and unduly stress a Western view of the world, they are of some interest. In 1904 H. J. Mackinder, a British geographer, drew attention to what he called the 'geographical pivot' of history (Map 2D), a vast landlocked region possessing substantial human and material resources, inaccessible to the maritime powers. Mackinder believed that, in time, the peoples of this region would reach out to control the marginal lands of Europe and Asia, ultimately dominating the Afro-Asian world-island, if not the world itself. The marginal lands of Eastern Europe and the Middle East were clearly of paramount importance to the maritime powers if this process was to be halted. Although Mackinder's views have been heavily criticised, and the extent of his influence on official thinking is in doubt, Western foreign policy since World War II has behaved as though the threat from the pivot area was real, as the Soviet Union has sought access to the maritime world. An American, N. J. Spykman, refined Mackinder's ideas in 1944 by stressing the fundamental role of the tier of states located on the margins of the pivot area, or 'heartland', where the struggle for influence would be crucial. Spykman called this the 'rimland', where some states would be aligned with the landpower, and others with the maritime powers. There were echoes of this theme in S. B. Cohen's 1964 scheme of world geostrategic regions, in which he introduced the concept of 'shatterbelts', one in the Middle East and one in Southeast Asia (Map 2D). Cohen's shatterbelts are seen as large, strategically located regions occupied by a number of rival states in which great power competition is fostered by the large number of political units and conflicting local interests.

Physically, the Middle East and North Africa are located in a region of great geological significance with regard to plate tectonics. The continental land masses are embedded in massive plates which move across the surface of the globe in response to convection currents deep in the earth. In the Middle East several such plates are in close proximity. The coming together of plates largely explains the presence of impressive fold mountains (Map 4) especially in northern parts, and their drifting apart accounts for the Red Sea and associated rift valleys. Zones of contact between plates are invariably the location of earthquakes, for which much of the region is notorious (Map 13). Maps 8–10 depict various climatic characteristics of the region, some of which are also largely attributable to global location in relation to atmospheric circulation and wind systems in the northern hemisphere.

KEY REFERENCES

Beaumont, P., Blake, G. H. and Wagstaff, J. M., *The Middle East: A Geographical Study* (Wiley, Chichester, 1976).
Chaliand, G. and Rageau, J. P., *Atlas Stratégique* (Fayard, Paris, 1983).
Drysdale, A. D. and Blake, G. H., *The Middle East and North Africa: A Political Geography* (Oxford University Press, New York, 1985).
Fisher, W. B., *The Middle East*, 7th edn (Methuen, London, 1978).
Kidron, M. and Segal, R., *The New State of the World Atlas* (Pan Books, London, 1985).

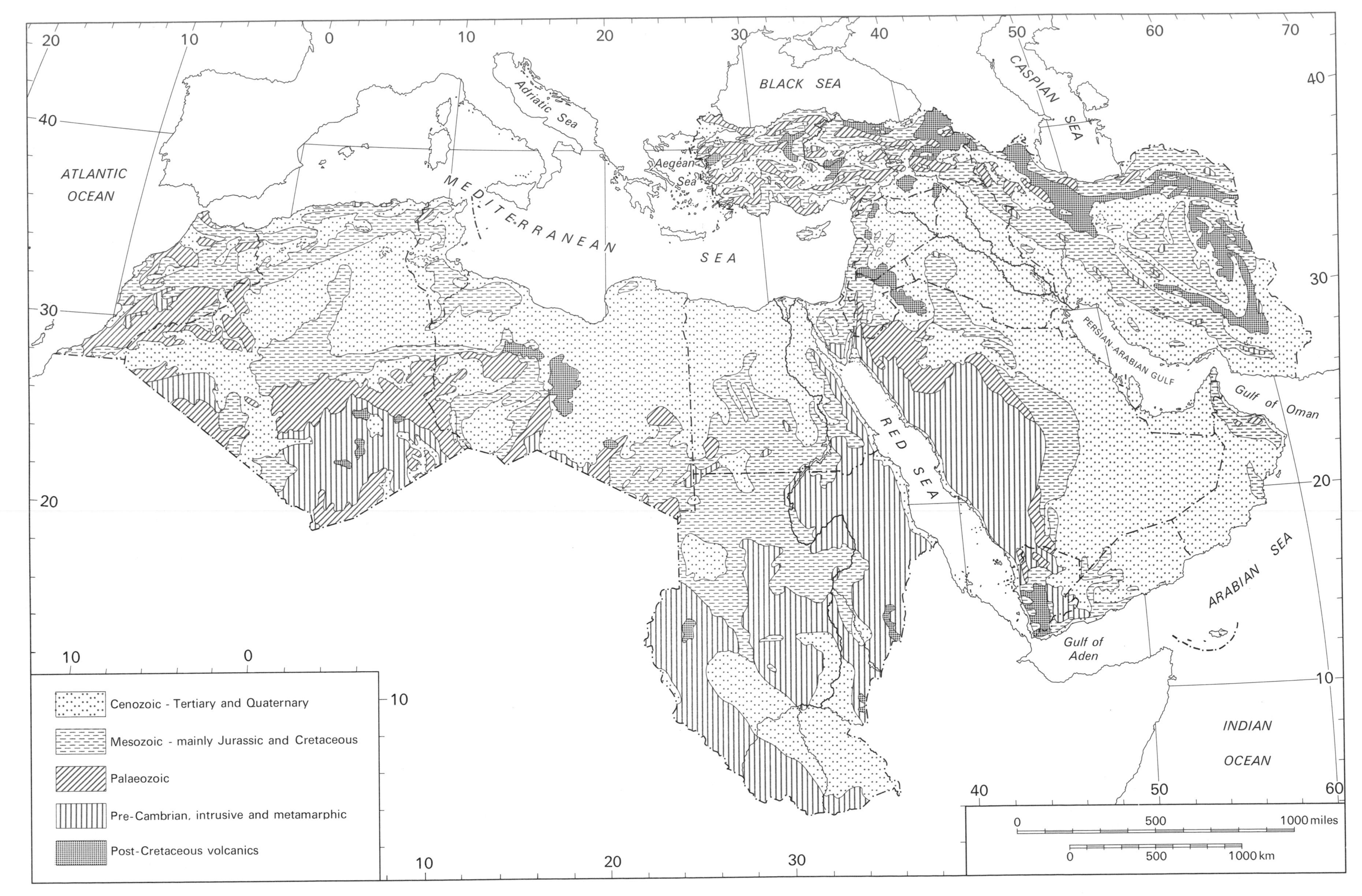

ATLANTIC OCEAN
MEDITERRANEAN
SEA
Adriatic Sea
Aegean Sea
BLACK SEA
CASPIAN SEA
PERSIAN-ARABIAN GULF
Gulf of Oman
RED SEA
ARABIAN SEA
Gulf of Aden
INDIAN OCEAN
Cenozoic - Tertiary and Quaternary
Mesozoic - mainly Jurassic and Cretaceous
Palaeozoic
Pre-Cambrian, intrusive and metamarphic
Post-Cretaceous volcanics
0
500
1000 miles
0
500
1000 km

3 Geology

Structure and relief

John Dewdney

The extremely varied relief of the countries of the Middle East and North Africa illustrated in Map 4 at first sight appears to bear little relation to the pattern of geological outcrops shown in Map 3; this is because a third factor must be taken into account, namely geological structure, the main elements of which are shown in Fig. 2. The relationships between structure, outcrop and relief are best understood if we recall the main features of the structure of the earth as a whole. Basically, there are two major elements, the stable blocks or 'continental platforms' and the intervening 'orogenic belts' or zones of mountain building. The continental platforms are composed of extremely tough igneous and metamorphic rocks, which were formed at great depth beneath the surface during mountain-building episodes in very early (Archaean or Pre-Cambrian) geological times. These hard materials have proved resistant to all later fold movements and, although they have been faulted, raised and lowered *en masse* by epeirogenic earth movements and have been subjected to long periods of sub-aerial denudation, their structures are very different from those of the intervening orogenic belts where, at various times, great thicknesses of sedimentary rocks derived from the denudation of the adjacent platforms have been thrown into folds by vigorous lateral orogenic or mountain-building earth movements, brought about by the horizontal pressure exerted by the movement of the platforms or tectonic 'plates'.

In the case of the Middle East and North Africa, we are concerned with two major blocks: the Afro-Arabian to the south and the European to the north, the latter, as it happens, lying wholly outside the region. Between the two there developed a major depression in the earth's crust, occupied by the Tethys Sea, into which vast quantities of sediment were poured as the two platforms were worn down by erosion. At a later stage, the Afro-Arabian plate moved northwards, throwing up massive fold mountain ranges – of which today's mountains are greatly reduced remnants – as it converged with the European plate.

In detail, the situation was much more complex than this simplified description implies. As movement occurred, the southern plate was split into three sections – the African, Arabian and Somalian plates – which are now separated by the deep down-faulted troughs occupied by the Dead Sea, the Red Sea and the great rift valleys of Ethiopia and Kenya. In addition, the northern edges of the plate were shattered, producing separate smaller plates beneath the Aegean and Turkey, giving particularly complex structures in the northern part of the region, where plateaux developed on small block fragments are surrounded by complex systems of fold mountain ranges.

As already indicated, the pattern of rock outcrops displayed in Map 3 gives little indication of these underlying structural features. Apart from the fact that, by definition, Archaean rocks can occur only in the platform areas, there is no clear distinction to be seen between the two zones in terms of the age of the rocks. Younger materials – in this case mainly Mesozoic, Cretaceous and Quaternary – occur in orogenic belts and on continental platforms alike. Given the small scale of the map and the fact that only two colours are used, only the major geological areas can be distinguished here. More detailed multicolour geological maps distinguishing individual rock systems show greater contrasts: whereas across the continental platforms a single outcrop may cover several hundreds of square kilometres, in the folded zones rapid changes of rock type across short distances are characteristic. This contrast reflects the fact that in platform areas the Archaean basement is for the most part covered by varying thicknesses of a wide range of sedimentary rocks but, because these sediments are underlain by a rigid basement which has protected them from later folding, they lie horizontally or have only shallow angles of dip. In the orogenic belts, on the other hand, dissection and faulting of tightly folded sediments with steep angles of dip produce rapid alterations of outcrop so that many different rock types appear at the surface within small areas of very varied relief. With these aspects of geological structure in mind, we can turn to a more detailed examination of the geology and relief of the region.

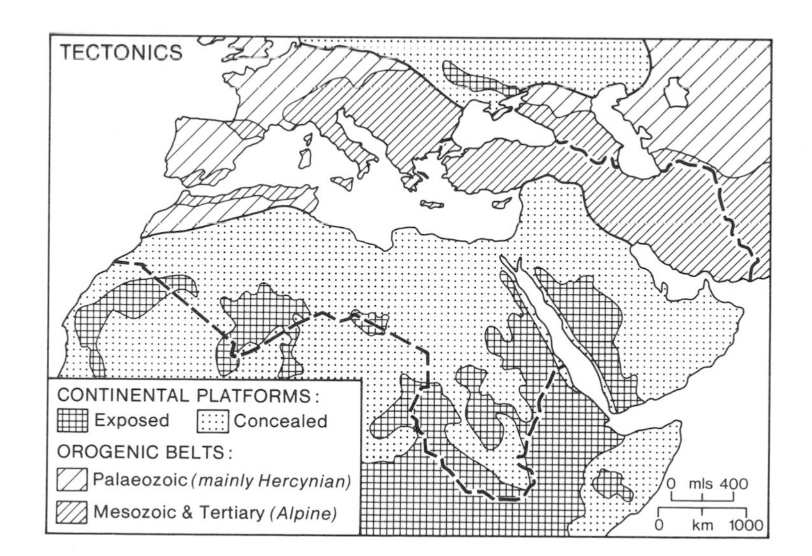

Fig. 2 Tectonics.

The largest single structural unit is the African platform, which occupies the southern parts of Morocco, Algeria and Tunisia and the whole of Libya, Egypt and Sudan. The Archaean basement outcrops over about a quarter of this area, generally producing dissected plateaux, which rise quite gently from the surrounding plains but reach considerable heights above sea level. One of the largest such areas is the Hoggar Massif of southern Algeria, with a maximum height of 3003 m; others are the Tibesti Plateau on the southern borders of Libya (which reaches 3265 m in northern Chad), and the Gilf Kebir Plateau (1250 m) in southwestern Egypt. Extensive exposures of the basement also occur in the Sudan, notably the Darfur Plateau (3088 m) in the west and the hills in the extreme south, which reach 3200 m along the border with Uganda. The Red Sea Hills between the valley of the Nile and the eastern coasts of Egypt and Sudan are of similar origin and top 2500 m in places; these are bounded on the east by very steep fault-line scarps overlooking the Red Sea. Between these upland massifs and the Mediterranean are vast expanses of gently sloping plains, developed mainly on Jurassic and Tertiary rocks but also carrying extensive superficial deposits of Quaternary age. While much of this area is less than 200 m above sea level, there are also occasional uplands, notably the Jebel Akhdar of eastern Libya, which reaches 865 m. The most striking features of this zone are the stony desert plains or hamadas and the major sand deserts or ergs. Further diversity is added by patches of salt marsh such as the chotts of southern Tunisia and the Qattara depression in western Egypt. In the latter case, wind erosion has excavated the surface to 133 m below sea level.

The Arabian platform extends northwards from the Arabian peninsula to underlie Israel, Lebanon, Syria, southeastern Turkey and much of Iraq. The basement rocks outcrop over a considerable area in western Saudi Arabia and give rise to an upland which reaches its greatest heights near the Red Sea coast. North of Mecca the crest line is broken, with maximum heights between 1500 and 2500 m. Further south, and occupying most of the Yemen Arab Republic, are the Asir Mountains, which reach 2500 to 3600 m. This upland area is bounded on its western and southern sides by fault-line scarps overlooking the Red Sea and the Gulf of Aden. Elsewhere, the Arabian platform is overlain by sedimentary rocks of Jurassic, Tertiary and Quaternary age, which form extensive plains sloping gently northeastwards to the valleys of the Tigris and Euphrates and the Arabian/Persian Gulf. As in North Africa, there are vast expanses of sand desert, of which the largest is the Rub al-Khali (the 'empty quarter') in southern Saudi Arabia. Covering an area in excess of half a million square kilometres, this is the largest sand desert in the Middle East,

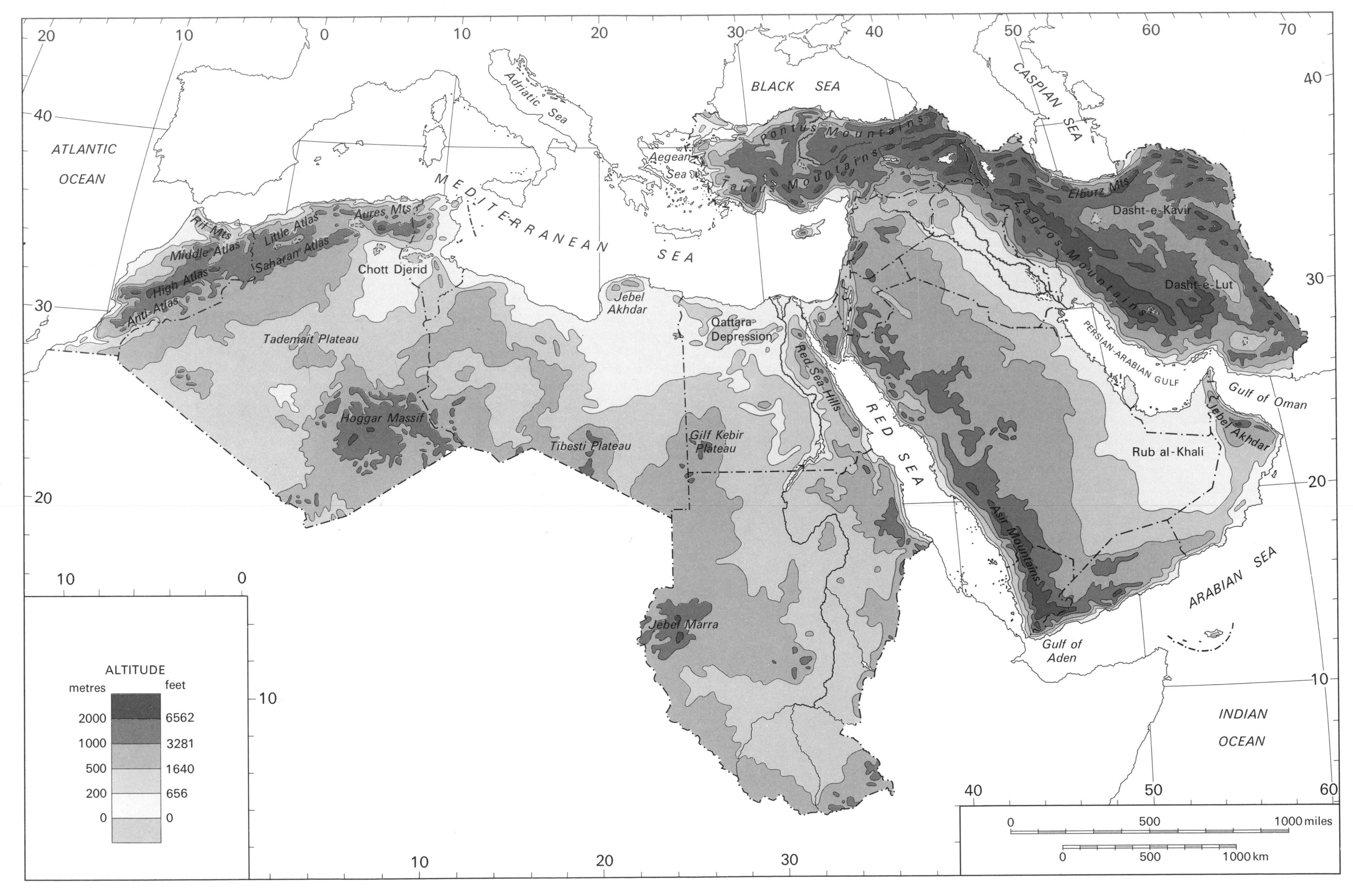

ATLANTIC OCEAN
MEDITERRANEAN SEA
Adriatic Sea
Aegean Sea
BLACK SEA
CASPIAN SEA
Pontus Mountains
Taurus Mountains
Elburz Mts
Zagros Mountains
Dasht-e-Kavir
Dasht-e-Lut
Rif Mts
Middle Atlas
Little Atlas
Aures Mts
Saharan Atlas
High Atlas
Anti-Atlas
Chott Djerid
Tademait Plateau
Hoggar Massif
Tibesti Plateau
Jebel Akhdar
Qattara Depression
Red Sea Hills
Gilf Kebir Plateau
RED SEA
PERSIAN-ARABIAN GULF
Gulf of Oman
Jebel Akhdar
Rub al-Khali
Asir Mountains
ARABIAN SEA
Gulf of Aden
Jebel Marra
INDIAN OCEAN
ALTITUDE
metres
feet
2000
6562
1000
3281
500
1640
200
656
0
0
0
500
1000 miles
0
500
1000 km

4 Relief

with dunes more than 200 m in height. In the extreme east of the peninsula, the Jebel Akhdar mountains of Oman, developed on volcanic rocks, reach 3352 m.

In the Lebanon, Israel and western Jordan, local relief is influenced by the presence of a faulted trough running north–south from the Jordan valley through the Dead Sea to the Gulf of Aqaba. The floor of the Dead Sea basin is more than 300 m below sea level; the mountains on either side of the rift valley reach above 2000 m in places.

The northern part of the region comprises a complex series of young fold mountains, mainly of Tertiary age, which occupy the northern parts of Morocco, Algeria and Tunisia, virtually the whole of Turkey and Iran and the eastern fringes of Iraq. These form part of the main belt of Alpine folding which extends from the Atlantic across southern Europe, North Africa and the Middle East to the Himalayas and beyond. There are numerous distinct mountain ranges, which at times converge in 'knots' and between which there are often high-standing plateau areas.

In Morocco, there are three main west–east trending ranges; from north to south they are the Middle Atlas, the High Atlas and the Anti-Atlas, with maximum heights of 3290 m, 4162 m and 2531 m respectively. The Jebel Toubkal of the High Atlas is the highest point in North Africa. In northern Morocco, the Rif Mountains (2452 m) run close to the Mediterranean coast and continue a line of folding which crosses the Strait of Gibraltar from Spain. In Algeria there are two main ranges: the Little Atlas in the north and the Saharan Atlas in the south. These enclose an extensive high plateau at heights of 750–1500 m. These two ranges converge to form the Aures Mountains of eastern Algeria and Tunisia.

The fold mountain belt occurs again in Turkey, where there are two main systems – the Pontus and Taurus – each consisting of several ranges, backing the Black Sea and the Mediterranean respectively. Crest lines above 2000 m are common in both systems (heights tend to increase towards the east) and there are several peaks above 3500 m. Between the Pontic and Tauric systems is the so-called 'Anatolian Plateau', which comprises a number of high plains around 1000 m separated by mountain ranges reaching 1500 m or more. The two systems converge in eastern Turkey to produce a complex upland massif around Mount Ararat (5165 m). Eastward from Ararat into Iran, the mountains again divide to form the two major systems of the Elburz and Zagros Mountains. The Elburz Mountains are a narrow range immediately south of the Caspian Sea and contain the highest peak in the region – Mount Damavand (5610 m). The Zagros Mountains, a more complex system comprising several ranges and intervening basins, run southeast from the Ararat massif through western Iran and along the northern coast of the Gulf. Crest lines are generally at 2500–3000 m, but there are several peaks above 3500 m; the highest of these is Zard Kuh (4548 m) to the southwest of Esfahan. Finally, in eastern Iran, a series of rather smaller ranges with a general northwest to southeast trend, and reaching heights above 3000 m in places, close in the eastern borders of the region. Set between the Elburz, Zagros and East Iranian mountain systems is a large plateau area at heights of 500–1000 m, which contains two major basins of inland drainage. The more northerly, the Dasht-e-Kavir, contains the biggest salt desert in the region; the more southerly, the Dasht-e-Lut, is a rocky desert with extensive sand dunes.

The Middle East and North Africa as a whole remains a zone of tectonic instability in which plate movement, folding and faulting are still in progress. In the past, these movements were accompanied by extensive outpourings of volcanic material, and several of the highest peaks, including Mount Ararat, are volcanic in origin. Today there are no active volcanoes and tectonic instability is manifested chiefly in the occurrence of earthquakes (see Map 13). These are most common in the eastern part of the region, in Turkey and Iran, where the structure is especially complex, but have also occurred in the Maghreb in recent years.

KEY REFERENCES

Fisher, W. B., *The Middle East* (Methuen, London, 1978).

Beaumont, P., Blake, G. H. and Wagstaff, J. M., *The Middle East: A Geographical Study* (Wiley, Chichester, 1976).

Economist Intelligence Unit, *Oxford Regional Economic Atlas of the Middle East and North Africa* (Oxford University Press, London, 1960).

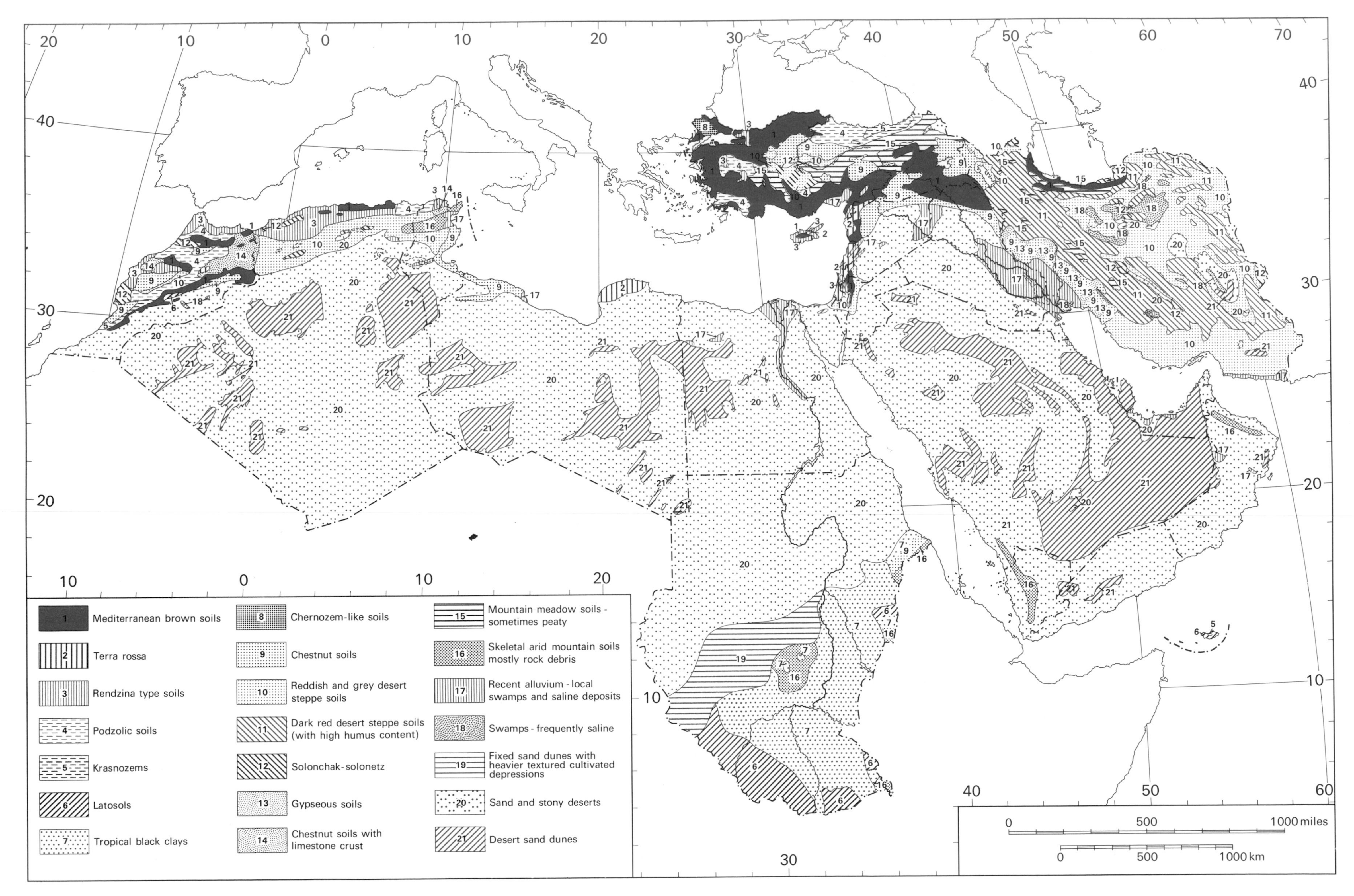

1 Mediterranean brown soils
2 Terra rossa
3 Rendzina type soils
4 Podzolic soils
5 Krasnozems
6 Latosols
7 Tropical black clays
8 Chernozem-like soils
9 Chestnut soils
10 Reddish and grey desert steppe soils
11 Dark red desert steppe soils (with high humus content)
12 Solonchak-solonetz
13 Gypseous soils
14 Chestnut soils with limestone crust
15 Mountain meadow soils - sometimes peaty
16 Skeletal arid mountain soils mostly rock debris
17 Recent alluvium - local swamps and saline deposits
18 Swamps - frequently saline
19 Fixed sand dunes with heavier textured cultivated depressions
20 Sand and stony deserts
21 Desert sand dunes
0 500 1000 miles
0 500 1000 km

5 Soils

Vegetation and soils

John Dewdney

It is well known that the patterns of soil and vegetation in any area are inter-connected, each affecting the nature of the other, and that both are closely related to the prevailing climate. The plants which constitute a vegetation assemblage will grow only within a certain range of temperature and moisture conditions and have specific demands for nutrients, which only particular types of soil can supply. At the same time, the soils in which the vegetation grows will themselves be affected by the nature of the plant cover, especially as regards their content of organic matter. Furthermore, the physical and chemical processes which produce a particular type of soil are strongly influenced by the climatic conditions under which that soil develops; indeed, the climatic factor is likely to be much more important than the nature of the parent material from which the soil is derived. In this section, soil and vegetation are dealt with in a single, integrated text but, for a fuller understanding of the climatic influence, the reader should refer to the various aspects of climate illustrated by subsequent maps.

Maps of the soils and vegetation types of an area as large as the Middle East and North Africa are inevitably somewhat complex, given the wide range of conditions within the region with regard to the various other elements of the physical environment. Climate, particularly temperature and rainfall as they affect the length of the growing season and the availability of moisture, is the dominant influence and varies with latitude, altitude and distance from the sea. At the local level, detailed variations in soil and vegetation are influenced by steepness of slope, exposure, drainage conditions and geology.

Inevitably, at the relatively small scale of the maps in this atlas, the soil and vegetation types shown must be limited in number and highly generalised, thus concealing much of the local detail so important to those who gain their living from the land. A further problem here is the number and variety of the classifications of soil and vegetation types devised by scholars in different countries, none of which can be regarded as definitive. Even when a classification system is decided upon, there are still difficulties. If a map is to be drawn, boundaries between types must be indicated, but in reality, soil and vegetation types only rarely, under very special conditions, change abruptly along a linear divide. Most commonly they grade, often imperceptibly, from one type to another through a broad transition zone; thus decisions have to be made not only as to where the boundary between two major types is to be placed but also by how many different subtypes the transition should be indicated. This is a particular difficulty in areas of gentle relief. In mountainous areas, on the other hand, an intricate mosaic of local soil or vegetation types covering small areas must be generalised to show the type predominant over the mountain zone as a whole.

In the discussion which follows, the method adopted is to concentrate attention on the somewhat simpler pattern presented by the vegetation map as the main aspect of regional variation and to relegate the discussion of soil types to a secondary position within that of each vegetation region.

The vegetation types fall readily into three main categories indicated by the three columns in the key to Map 6, with desert types at one extreme, forest types at the other and various kinds of scrub and grassland in between. In general terms, this threefold division is closely related to the pattern of annual precipitation shown in Map 9; forest types occur mainly in areas with annual rainfall in excess of 400 mm and desert types in areas with less than 50 mm, though local exceptions to these limits can, of course, be found.

Forest vegetation

Forest types are found mainly, but not entirely, in upland areas, covering most of Turkey (excluding the semi-arid interior and southeast), the Zagros, Elburz and eastern mountain ranges of Iran, forming a broken circle around the much drier interior of that country, Cyprus and the Levant coastal strip. In North Africa, forest is widespread in the northern Atlas of Tunisia and Algeria; in Morocco it also extends southwards into the High Atlas as a result of that country's heavier rainfall due to its proximity to the Atlantic. Some less elevated parts of North Africa are also allocated to the forest category: the Jebel Akhdar of Cyrenaica (eastern Libya) and the coastal lowlands of western Tripolitania and eastern Tunisia.

Care should be taken not to interpret the description 'forest' as in any sense indicating the presence of a continuous tree cover throughout the areas thus labelled. In the first place, as already indicated, the map is highly generalised and 'forest' zones are in fact a mosaic of forested and unforested areas. Secondly, the map attempts to show 'natural' vegetation, the indigenous vegetation type unaffected by human activity. Such 'natural' vegetation in reality covers only a fraction of the 'forest' zones and, over large areas, forest has either been cleared to provide land for agriculture or degraded as a result of timber cutting and/or uncontrolled grazing. Thus, in Turkey, for example, only about a quarter of the land area is officially classed as 'forest or woodland' and, of this, barely 20% (i.e. 5% of the land area) is 'productive forest'.

Five different types of forest are distinguished on the map, of which the most luxuriant is the deciduous forest found along the Black Sea coast of Turkey and between the Elburz and Caspian in Iran. In these areas, which have the heaviest rainfall, no summer drought and mild winters, there is a great variety of species. Sweet chestnut, hornbeam, oriental spruce, alder and oak are the most common species and there is a rich shrub layer of rhododendron, laurel, holly, myrtle, hazel and walnut.

A second type, labelled 'mixed forest', occurs in the rather drier areas of northwest, north-central and northeast Turkey and on the Elburz and Zagros ranges of Iran. Winters can be cold and there is a fairly pronounced summer drought so that the forest is less rich, more open and less continuous than in the humid deciduous zone. The main species are oak and juniper with a considerable admixture of pine and fir, these coniferous types becoming more common with increasing altitude. Contained within this zone are sizeable areas of open grassland. True coniferous forest, the third type, is confined to the Taurus ranges of southern Turkey.

Mediterranean maquis-forest forms a discontinuous belt around the Mediterranean basin, occurring in eastern and southern Turkey, along the Levant coast and over large areas of the Maghreb. The dominant influence here is the summer drought characteristic of the Mediterranean climate regime. At higher altitudes, the main species are pines, firs and oaks, with cedar, oriental beech and juniper also present. At lower elevations, true maquis occurs, dominated by myrtle, wild olive, laurel and carob with only occasional oak, pine and cypress. Within this general zone, a special feature is the cork oak forest of northern Morocco and northeastern Algeria.

As Map 5 indicates, there is a considerable variety of soil types within the forest zones, of which by far the most widespread is the Mediterranean brown soil (1), found particularly widely in Turkey. This soil experiences little leaching and, by the standards of the region at least, is relatively rich in humus. In areas developed on limestone, it is replaced by clayey terra rossa (2) or lime-rich rendzina (humus-carbonate) soils (3); in wetter areas – northeastern Turkey, for example – by leached, acidic, podzolic soils (4). Heavily leached red or yellow clay soils, the krasnozems (5), are confined to areas of very heavy rainfall. Gypseous soils (13), developed from gypsum rock, occur widely in the Zagros mountains.

Scrub and grassland vegetation

Between the forest types described above and the true desert are various scrub and grassland vegetation types; the distribution of these is closely related to annual precipitation, which is everywhere too low to support extensive tree growth. The most extensive of these are the

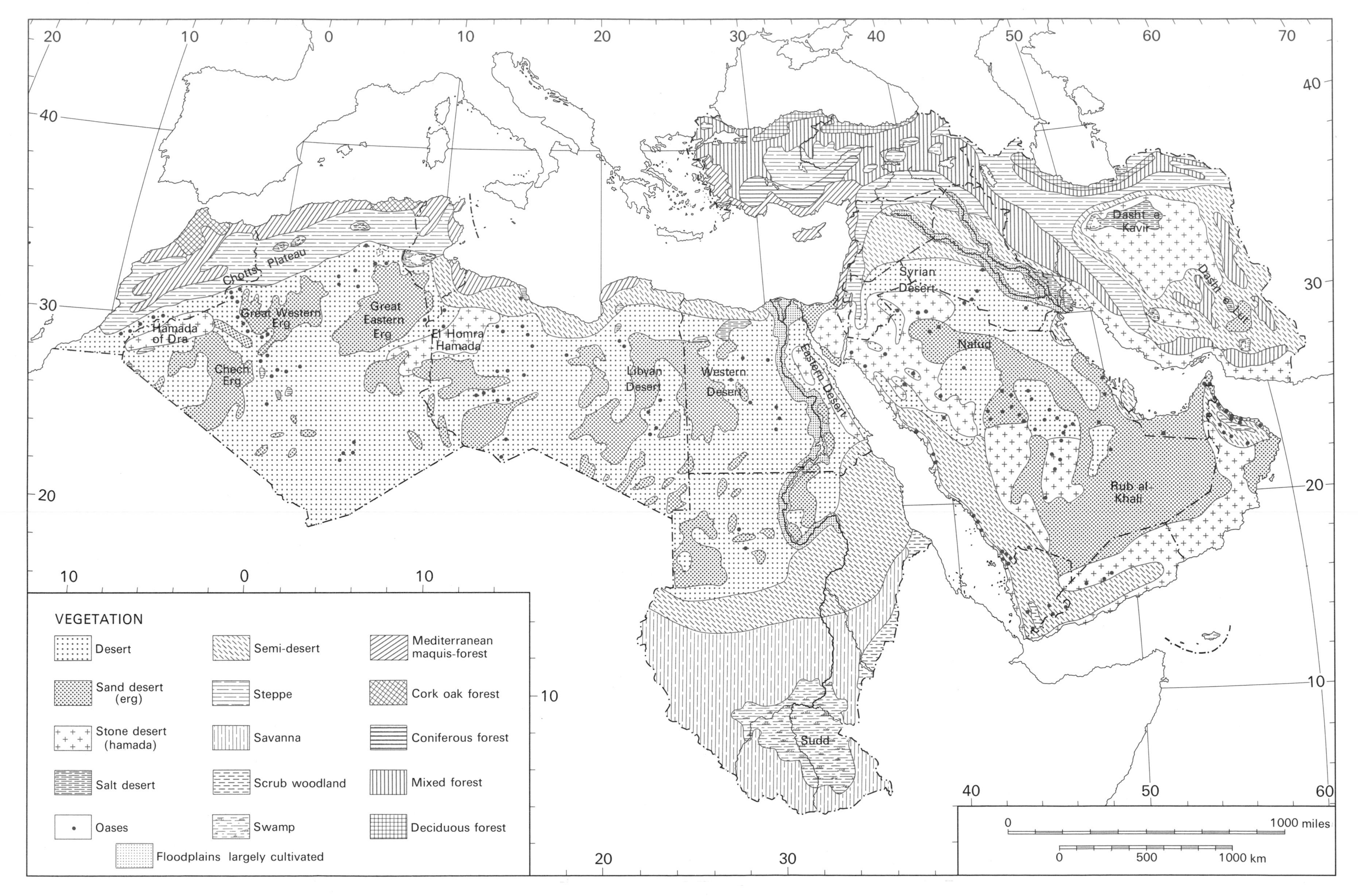

6 Vegetation

steppe and semi-desert types, which grade into each other as the rainfall diminishes and form a near-continuous belt right across the region on the northern side of the Sahara and Syrian deserts.

The steppe is distinguished from the semi-desert by its higher rainfall and somewhat more plentiful cover but even here the cover is sparse and largely confined to grasses and herbaceous shrubs. These grow rapidly in spring, the period of maximum rainfall, but are burnt off by the drought and high temperatures of the summer months. Only in the best-watered areas is a grass cover retained throughout the year. The steppe covers a broad belt right across the Maghreb, inland of the forest zone, and recurs in the dry central zone of Anatolia, in an arc from western Jordan and Syria through southern Turkey to northern Iraq, and in a semi-circle between the Dasht-e-Kavir and the Elburz and Zagros mountains of Iran. In North Africa and Iran the steppelands contain relatively little cultivated land and are devoted mainly to pastoralism, but in Turkey and northern Syria there is extensive, low-yielding cereal cultivation. Semi-desert occurs between the steppe and desert in Syria, Iraq and Iran and between the maquis zone and the desert in northern Libya and Egypt. Here the rainfall is even lower and the grassland vegetation yet more sparse. Over much of the semi-desert zone there is extensive bare land dotted with drought-resistant shrub species.

There is a great variety of soils in the steppeland areas. In the Maghreb steppe they are highly calcareous with a low organic content (desert steppe soils, 10) but in Iran they are a good deal richer in organic matter (dark red desert steppe soils, 11). Chestnut soils (9), which are intermediate between these two types, are common in Syria and Turkey. In the semi-desert zones, the soils differ from those of the true desert only in having a very slightly higher humus content.

On the southern side of the Sahara, in Sudan, the sequence of vegetation reflects a southward increase in both temperature and rainfall. In common with other interior areas of sub-Saharan Africa, this is a zone with pronounced dry and wet seasons and vegetation is adapted to withstand a long period of drought. The increase in the length of the rainy season from north to south is responsible for the sequence of vegetation types. Thus immediately south of the desert is a semi-desert zone with poor, intermittent grassland vegetation and some acacia, and this merges into the savanna, a mosaic of grassland and semi-deciduous woodland. Tree species become dominant in the southwest of the country, which is the wettest area.

There are three main types of soil. In the northwestern part of the savanna zone is an area of ancient sand dunes (19) formed during a dry phase of the Pleistocene which, under subsequent moister conditions, have been fixed by the development of savanna vegetation. Tropical black clays (7) cover most of the remainder of the savanna zone, but in the southwest, the wettest area, these are replaced by the more strongly leached and deeply weathered latosols (6) or tropical red earths and loams.

Deserts

Deserts cover at least half the surface of the region, including the great bulk of the territory of Algeria, Libya, Egypt and the Arabian peninsula as well as sizeable parts of Morocco, Sudan, Jordan, Iraq and Iran and small sections of Tunisia and Israel. Of all the countries of the Middle East, only Lebanon, Cyprus and Turkey have no desert territory. Map 6 indicates the main areas of sand and stone desert (erg and hamada respectively), but throughout the desert zone as a whole the vegetation is extremely sparse and in many sections non-existent. As a result, the humus content of the soils is very low, soil horizons are only poorly developed and the texture is coarse with many unweathered rock fragments. Scattered throughout the desert zone are numerous oases where depressions in the surface tap the water table to produce small pockets of intensive cultivation.

A number of special types of vegetation and associated local soils are also indicated. Salt deserts occur in basins with no drainage outlet. Water flowing in from surrounding areas in late winter and early spring lies on the surface to be evaporated during the summer, leaving salt encrustations on the surface of the land and saline deposits in the soil profile. The most extensive salt desert is to be found in the lower part of the Dasht-e-Kavir in Iran; similar conditions occur in the Qattara depression of northern Egypt and the 'chotts' of Algeria and Tunisia.

A major area of swamp vegetation occurs in the Sudd district of southern Sudan. Here, the waters of the upper Nile and its tributaries converge in a vast alluvial plain which is flooded in the rainy season but is a parched wilderness in the dry. Permanent swampland is confined to the immediate vicinity of the rivers.

Finally, the map delimits the major alluvial flood plains of the Nile, Tigris and Euphrates where any natural vegetation has long since been removed and replaced by irrigated agricultural land which contrasts vividly with the desert and semi-desert lands on either side.

KEY REFERENCES

Allan, J. A., 'Renewable natural resources' in J. I. Clarke and H. Bowen-Jones (eds.), *Change and Development in the Middle East* (Methuen, London, 1981).

Economist Intelligence Unit, *Oxford Regional Economic Atlas of the Middle East and North Africa* (Oxford University Press, London, 1960).

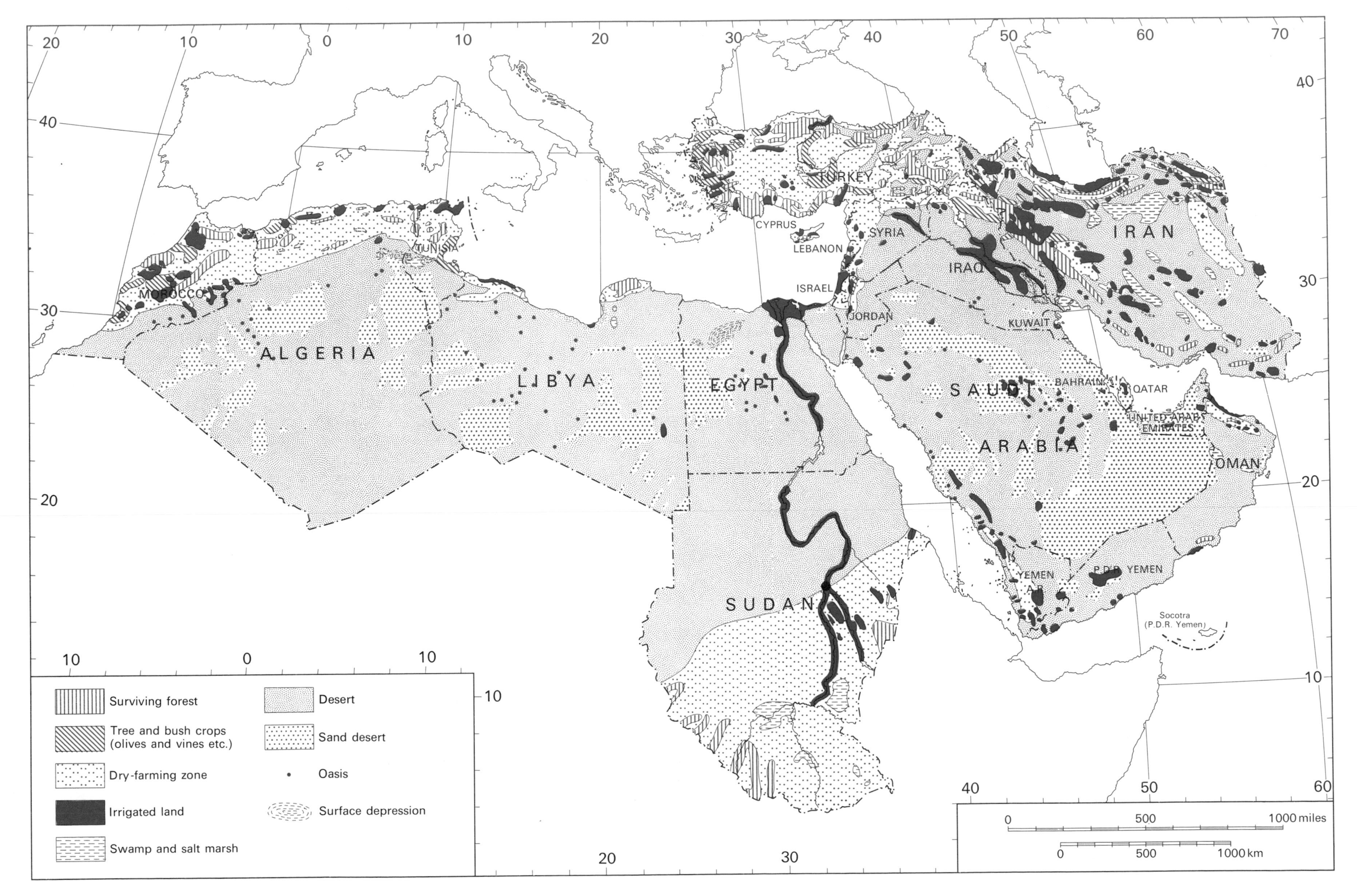

7 Landscapes

Landscapes

John Dewdney

The contemporary landscapes of the Middle East and North Africa, like those of any other region, have been developed as a product of human actions of various kinds operating against a background of physical environments which themselves differ very greatly from one part of the region to another. In some areas the physical environment is so forbidding and so inflexible that the natural elements of the landscape remain predominant and the impact of man has been so limited as to have brought about little or no change. On the other hand, in the most favourable areas, mankind has had such a massive impact on the local environment that little or nothing which is truly 'natural' has survived that impact.

Clearly one of the most important elements determining the extent to which landscapes have been transformed from the 'natural' to the man-made has been the patterns of population distribution and density illustrated by Maps 21 and 22 and discussed in the accompanying texts. Equally obvious is the point that here we are concerned with a two-way relationship. The potential of an area for human occupation and development is determined primarily by the nature of its physical environment, and those areas with relatively favourable environments offering the greatest potential for development by man – the alluvial floodplains of the great river valleys, for example – are the ones where concentrations of population have occurred and where the landscape has been most vigorously transformed. At the other extreme – in the great deserts of the region, for example – the natural environment has offered so little to mankind that population numbers and densities have remained very low and thus the modification of the natural environment has been minimal.

These are, of course, very general statements and take no account of the great variety of human conditions among the 22 countries of the region. According to the levels of development achieved and the nature of the national economies, man modifies the natural environment in different ways and to varying degrees. A physical environment of a particular type may be intensively developed in a small country with high population density where land resources are limited and population growth places increasing pressure on those resources, while an area of similar environment in a larger, less densely populated country with more generous land resources may be completely undeveloped. Equally, a country with large capital resources, which may themselves be derived from some specific element in the environment (oil is an obvious example), can undertake the development of environments which may remain completely undeveloped in a poorer country.

It follows, then, that our map of landscapes is a composite one containing a variety of elements which are mapped individually or in systematic classifications elsewhere in this atlas. In particular, it repeats elements shown on the maps of soils (5), vegetation (6) and agricultural regions (28). This reflects that, throughout virtually the whole of the Middle East and North Africa until quite recently, the most important aspect has been the potential offered by the environment for agricultural activities.

With the obvious exceptions of the urban and rural settlements and the (mainly recent) industrial sites of the region, where the land has disappeared beneath buildings and other structures and the landscape is completely man-made, the greatest modifications to the natural environment have occurred in the irrigated lands. Here, too, the landscape is almost entirely man-made; an intricate network of small fields and irrigation channels cover these lands which are the most intensively utilised lands of the region, supporting the highest population densities and providing the greatest return per unit area. On a small-scale map, irrigated areas can be shown only in the most general way, and distinguishing between irrigated and unirrigated lands at the national and local scales presents problems. Nevertheless, a distinction can be made between those countries in which agriculture depends almost entirely on irrigation and those where irrigated and dry-farmed areas form a complex mosaic. Most clearly in the former category is Egypt, where practically all farmland is irrigated and its distribution corresponds to the floodplain and delta of the Nile. In the countries of the Arabian peninsula, too, the great bulk of the arable land is irrigated, the only exceptions being in the uplands of Yemen A.R. and Oman. In a number of other countries, although there are quite extensive dry-farmed areas, the irrigated lands support the bulk of agricultural production. This applies in northern and central Sudan, where the Nile valley irrigated zone continues, and in Iraq, where the valleys of the Tigris and Euphrates are the main agricultural areas. In Iran, on the other hand, where irrigated land makes up about a quarter of the arable area, the pattern is one of a large number of separate blocks, most of them in piedmont zones between the mountains and the semi-arid and arid plains.

In the remaining countries of the region, the irrigated lands are a much smaller proportion of the total and occur in a variety of situations: along the main river valleys (though none of these is comparable to the Nile or Mesopotamian regions), as in Turkey; in basins in upland and mountain areas, as in Morocco and Algeria; or in coastal lowlands as in Tunisia, Libya and Israel.

In areal terms, though not in terms of production, the most important land use category shown on the map is the dry-farming zone where generalised arable farming with cereals as the dominant crops, interspersed with grazing areas, is the main form of land use. Comparison with Map 6 shows that dry farming corresponds mainly with the steppe grassland and savanna zones, though in the northern part of the region, particularly in Turkey, dry farming also extends into formerly forested areas. Here, too, over large areas, the landscape has been completely transformed by human activity. The natural grassland and, less commonly, woodland have been replaced by arable characterised, in contrast to the irrigated areas, by large open fields and a relatively thin scatter of village settlements. During fallow periods and in grazing areas there is some resemblance to the natural landscape but the plant species on fallow and grazing lands are usually fewer and often less luxuriant than those of the natural vegetation.

Set mainly within the dry-farming zones are considerable areas in which tree crops are predominant, though in detail such areas are mosaics of tree-crop and field cultivation. The landscapes of the tree-crop areas are also essentially man-made. Cultivated olives, vines, figs, nuts and fruits of many kinds, often planted in regular rows, are clearly distinguished from other types of vegetation.

Landscapes dominated by nature rather than by man cover a very large part of the region. This description can certainly be applied to the vast desert zones but is less realistic for the small areas of 'surviving forest' in which, even though a forest cover remains, it is likely to have been modified by human action and by the depredations of grazing animals. In any case, as comparison with Map 6 will show, such areas are only a small fraction of the original forest.

KEY REFERENCES

Beaumont, P., Blake, G. H. and Wagstaff, J. M., *The Middle East: A Geographical Study* (Wiley, Chichester, 1976).

Despois, J., 'Development of land use in Northern Africa and evolution of land use in South-Western Asia' in L. D. Stamp (ed.), *A History of Land Use in Arid Regions: Arid Zone Research A* (U.N.E.S.C.O., New York, 1961).

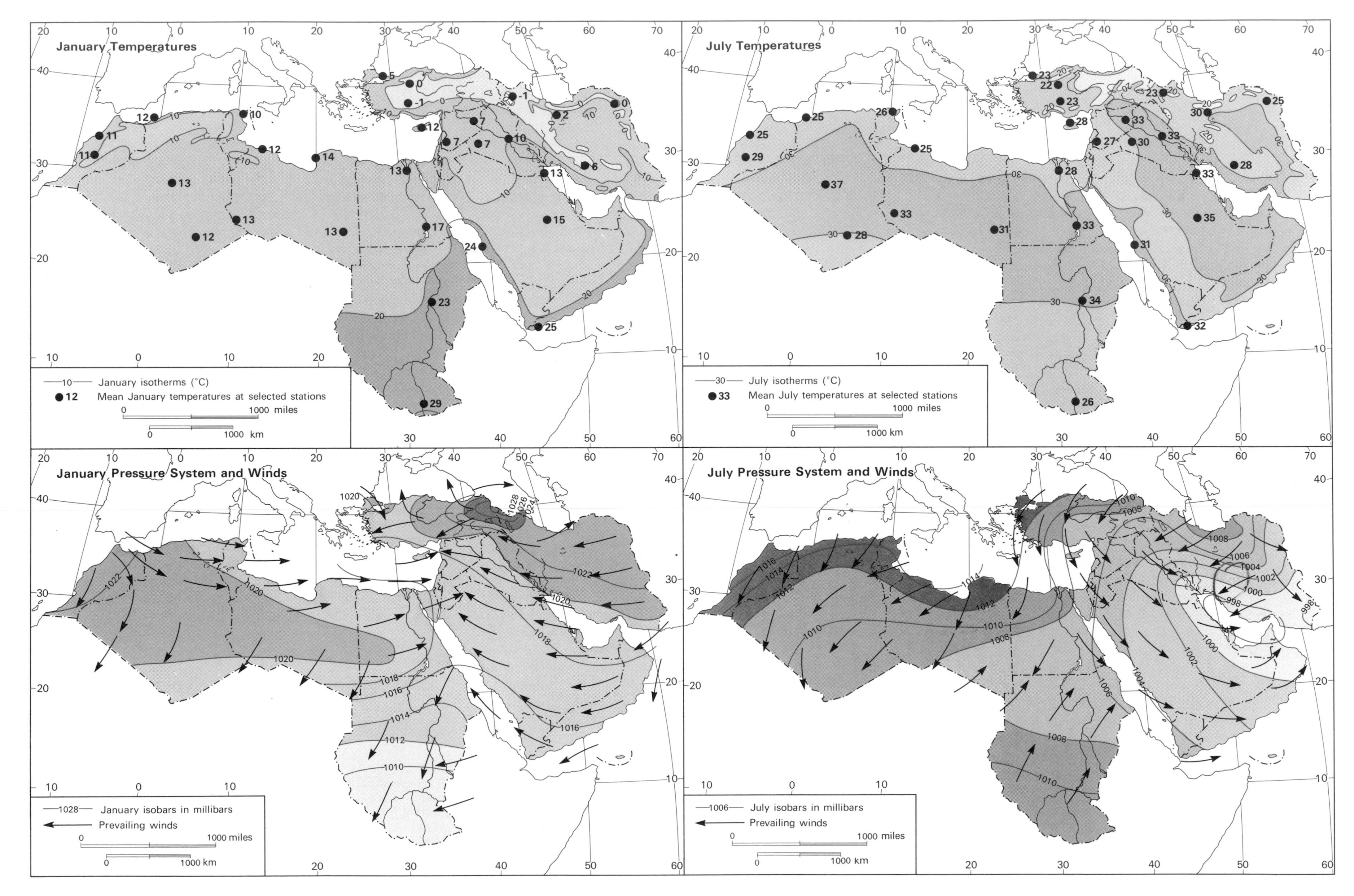

8 Temperature, pressure and wind direction

Temperature, pressure and wind direction

John Dewdney

Average January and July values for the three closely related climatic elements of temperature, pressure and wind direction are shown together in the four sections of Map 8. It is important to realise that the data displayed here are averages only and that actual conditions at particular times can diverge widely from these mean values. Among the factors responsible for the patterns shown are latitude, altitude and distance from the sea (continentality). These produce major regional contrasts, but conditions at individual climatic stations are likely to be affected by local factors, such as exposure and relief, which can upset the broader patterns.

As regards latitudinal influences, it is possible on a world scale to distinguish a well-developed high pressure system around latitude 30°N from which air moves out towards the zones of low pressure around 60°N and the equator. The rotation of the earth deflects the moving air so that the flow is predominantly from southwest to northeast (the mid-latitude westerlies) to the north of the high pressure zone and northeast to southwest (the northeast trade winds) to the south. Owing to the tilt of the earth's axis and the rotation of the earth around the sun, these wind and pressure belts are further north in summer than in winter and this gives much of the region a climatic regime in which the dry northeast trades are predominant in the summer months and the moist Atlantic southwesterlies in the winter (i.e. the Mediterranean type of climate with its winter rainfall maximum and summer aridity).

Some hint of this arrangement can be seen in the maps, but there are numerous local anomalies due to differences in altitude and in distance from the sea. Owing to the differential heating of land and sea, there is a general tendency for pressure to be higher over the land than over the sea during the winter months and vice versa in the summer. This effect can be seen, for example, on the map of January pressure and winds, which indicates a pronounced high pressure zone in the mountainous area of the Caucasus at the junction of Iran, Turkey and the U.S.S.R. In addition, the concept of winter low pressure in Mediterranean latitudes refers to average conditions; in reality it is the product of the passage from west to east across the area of a succession of low pressure cells ('depressions' or 'cyclones') which, in their passage, draw in winds from all directions.

The situation is perhaps simplest in the summer months, as shown by the map of July pressure systems and winds. An intense low pressure, associated with mean temperatures above 30 °C (86 °F), over the Gulf and the adjacent lowlands has a profound effect on wind directions in the eastern half of the region, causing a northwest–southeast (rather than northeast–southwest) movement over the Arabian peninsula, for example. Over most of North Africa, on the other hand, the 'normal' northeast trades can be seen. Southern Sudan, however, lies south of the subtropical low pressure belt and experiences winds from the southwest. All these winds are dry, originating as they do over the land.

In the winter season pressure is high over the Eurasian land mass to the north of the region and there is a relatively weak ridge of high pressure over the Sahara. Between these two high pressure areas is a zone of relatively low pressure in the Mediterranean basin, the prevailing westerly winds in this zone being the strongest component in the anti-clockwise wind circulation associated with the passage of strings of depressions. Pressure is at its lowest in the eastern Mediterranean and highest in the Caucasus and northern Iran, features which account for the dominance of easterly winds over Iran and the Arabian peninsula.

Average temperatures in July show relatively little variation. The hottest areas, recording means of more than 30 °C (86 °F), are in the deserts of North Africa and the Arabian peninsula and in the Tigris–Euphrates lowlands. It is somewhat cooler (around 25 °C (77 °F)) along the Mediterranean coast and the only areas below 20 °C (68 °F) are in the high mountain areas of eastern Turkey and northwestern Iran.

January temperatures show a pronounced latitudinal gradient from only 5 °C (41 °F) at Istanbul to nearly 30 °C (86 °F) in the extreme south of Sudan, modified by the effects of altitude, which produce a sizeable area with January means below freezing in Anatolia and the mountains of Iran. The range of temperatures is clearly much greater in the winter than in the summer, as is shown by Table 2.

Table 2. *Mean monthly temperatures (°C)*

	J	F	M	A	M	J	J	A	S	O	N	D	Year	Range
Casablanca	12	12	14	15	17	20	23	23	22	19	16	13	17	11
Alger	12	12	14	16	18	22	24	25	23	20	16	13	18	13
Tunis	10	11	13	16	19	23	26	26	24	20	15	12	18	16
Tripoli	12	13	16	18	21	23	25	27	26	23	18	14	20	15
Tamanrasset	12	14	18	22	26	28	28	28	26	23	18	14	22	16
Cairo	13	13	16	20	24	27	28	27	25	22	18	14	21	15
Khartoum	23	25	28	32	33	33	34	31	32	32	28	25	29	11
Juba	29	29	29	29	27	27	26	26	27	27	28	28	28	3
Izmir	8	8	12	15	20	24	27	27	23	19	14	10	17	19
Erzurum	–9	–7	–3	5	11	15	19	20	15	9	2	–5	5	29
Tehran	2	5	9	16	21	27	29	30	25	18	12	6	17	28
Abadan	12	15	18	24	31	34	35	35	32	27	21	14	25	23
Tel Aviv	14	14	17	19	23	26	27	28	27	24	21	16	22	14
Damascus	7	8	12	15	21	26	27	26	22	18	13	8	17	20
Baghdad	10	11	16	22	28	32	33	34	31	24	17	11	23	24
Aden	25	26	27	30	32	33	32	31	32	29	28	27	29	8
Range	38	36	32	27	22	19	16	15	17	23	26	33	24	26

Source: Air Ministry, Meteorological Office, *Tables of Temperature, Relative Humidity and Precipitation for the World* (H.M.S.O., London, 1958).

KEY REFERENCES

Beaumont, P., Blake, G. H. and Wagstaff, J. M., *The Middle East: A Geographical Study* (Wiley, Chichester, 1976), ch. 2, 'Climate and water balance'.

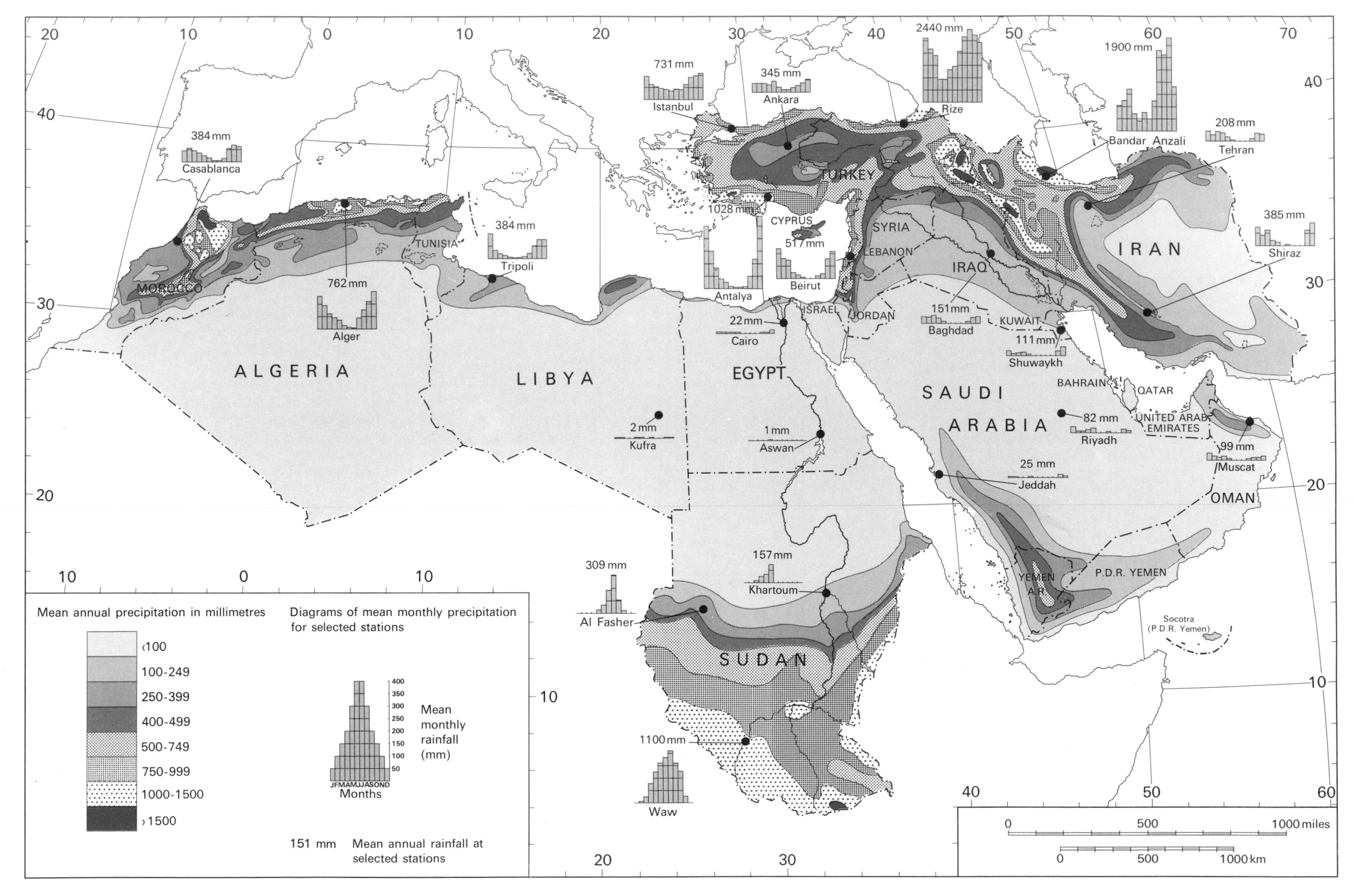

9 Precipitation

Precipitation

John Dewdney

Of all the elements of the physical environment of the Middle East and North Africa, none is more important than the annual rainfall, and none shows greater variation, ranging from well below 100 mm in the Saharan and Arabian deserts to nearly 2500 mm along parts of the Black Sea coast. Over very large parts of the region, temperatures (see Map 8) are such that plant growth can continue throughout the year; in such a situation it is the availability of water which is the most important climatic control on agricultural and other human activities.

Precipitation takes place when air containing water vapour is lifted to a height at which condensation occurs, and this may result from the mixing of air masses along fronts (cyclonic precipitation), from the presence of a mountain barrier (orographic precipitation), or from intense heating of the land causing the air above it to rise (convectional precipitation). Within the region, cyclonic disturbances move from west to east throughout much of the year, and there are major mountain systems aligned in several different directions (see Map 4) and extensive plains and plateaux where convection takes place. Consequently, all three types of precipitation contribute to the total, cyclonic and orographic, often in combination, being the more important.

Thus the general pattern of annual rainfall is a fairly simple one, in which the total diminishes from north to south towards the desert belt, increasing again to the south of the Sahara in central and southern Sudan. The rainfall received at any particular place, however, is influenced not only by its latitude but also by its altitude, its exposure to the prevailing wind, its distance from a major moisture source and its location with respect to the tracks followed by depressions; consequently, the detailed pattern is quite complex, with striking changes over short distances.

Areas with an average annual precipitation of more than 750 mm, an amount which can be described as 'abundant' by Middle Eastern standards, are limited in extent and, for the most part, coastal. Wettest of all are the Black Sea coastal zone of Turkey, where values range from 730 mm in the west (Istanbul) to more than 2400 mm in the east (Rize), and the Caspian coastlands of Iran (Bandar Anzali, 1900 mm). In both these areas, the passage of depressions at all seasons ensures the absence of any marked dry period, an unusual situation in the Middle East. Although, even here, there is a pronounced winter maximum with several monthly averages in excess of 250 mm and equivalent to the annual total at many Middle Eastern stations, summer months with less than 100 mm are rare. Several other coastal areas have totals of 750–1000 mm, including those of southern Turkey, the Levant and the Maghreb, but in these cases the summer drought characteristic of the Mediterranean climate occurs.

In all these coastal areas, precipitation increases rapidly in the mountains which back the sea, but then declines dramatically in the rain-shadow areas further inland.

There are two other zones to be placed in the heavy rainfall category. The first of these is the mountain zone running northwest–southeast from Lake Van in Turkey into the Zagros Mountains of Iran – an area of great importance in the water budget of that country, and a prime example of the influence of relief. The second is in the southern Sudan. Here, in marked contrast to the rest of the region, there is a strong 'summer' maximum and a 'winter' drought, this area lying at the northern edge of the African monsoon belt and receiving its moisture mainly from the southwest.

Areas of moderate annual rainfall between 500 and 700 mm include most of Turkey, Lebanon – the only country in the region which lacks extensive arid areas – most of northern and western Iran, and the Maghreb as far inland as the southern slopes of the Atlas Mountains.

Beyond these relatively well-watered areas there is a very abrupt decline in annual precipitation towards the deserts of North Africa, Arabia and eastern Iran. Tehran, for example, which is situated about 100 km south of the Caspian, but on the dry side of the Elburz Mountains, has just over 200 mm; Damascus and Amman, both about 100 km from the Levant coast, have 218 mm and 280 mm respectively. Another 100 km to the east the average is less than 150 mm.

Low rainfall totals occur further south without the intervention of a mountain barrier. In Libya, for example, the coastal capital of Tripoli receives rather less than 400 mm, Benghazi about 250 mm, and in Egypt Alexandria has 175 mm and Cairo only 22 mm. Conversely, in the Arabian peninsula, the great bulk of which receives less than 100 mm, the presence of mountains in the southwest and southeast produce totals above 500 mm in Yemen A.R. and over 350 mm in the Jebel Akhdar of Oman.

The uneven distribution of precipitation over the land areas of the Middle East and North Africa is a major influence on the agricultural potential of the various countries and on the distribution of cultivated land. The reader will also notice some correspondence between the rainfall map and that of population density (Map 22) – though with major anomalies where agriculture is supported by the irrigation of desert lands, as in the valleys of the Nile, Tigris and Euphrates.

A feature not illustrated by the rainfall map is the year to year variability of precipitation, which becomes more marked as the average diminishes. Cairo, for example, with an annual average of 22 mm, has experienced years with as little as 1.5 mm and as much as 64 mm; at Bahrain the mean is 76 mm and the range from 10 mm to 169 mm.

Throughout this discussion, the words 'precipitation' and 'rainfall' have been used interchangeably, but snow is by no means uncommon in the region and can occur at sea level almost anywhere north of latitude 30°, that is throughout the northern Maghreb, the Levant, Turkey and Iran. The depth and duration of snow cover depend, of course, on latitude and altitude. On the higher ranges of the Atlas, Taurus, Elburz and Zagros mountains it may lie for as long as six months, and the plateau areas of Turkey and Iran, together with mountain districts of Cyprus and Lebanon, have a snow cover lasting six to eight weeks. Winter storage of snow in mountain areas and its release as meltwater in spring and early summer are important elements in the water balance of several areas and the main support of irrigation systems in several parts of the region.

KEY REFERENCES

Beaumont, P., Blake, G. H. and Wagstaff, J. M., *The Middle East: A Geographical Study* (Wiley, Chichester, 1976), ch. 2, 'Climate and water resources'.

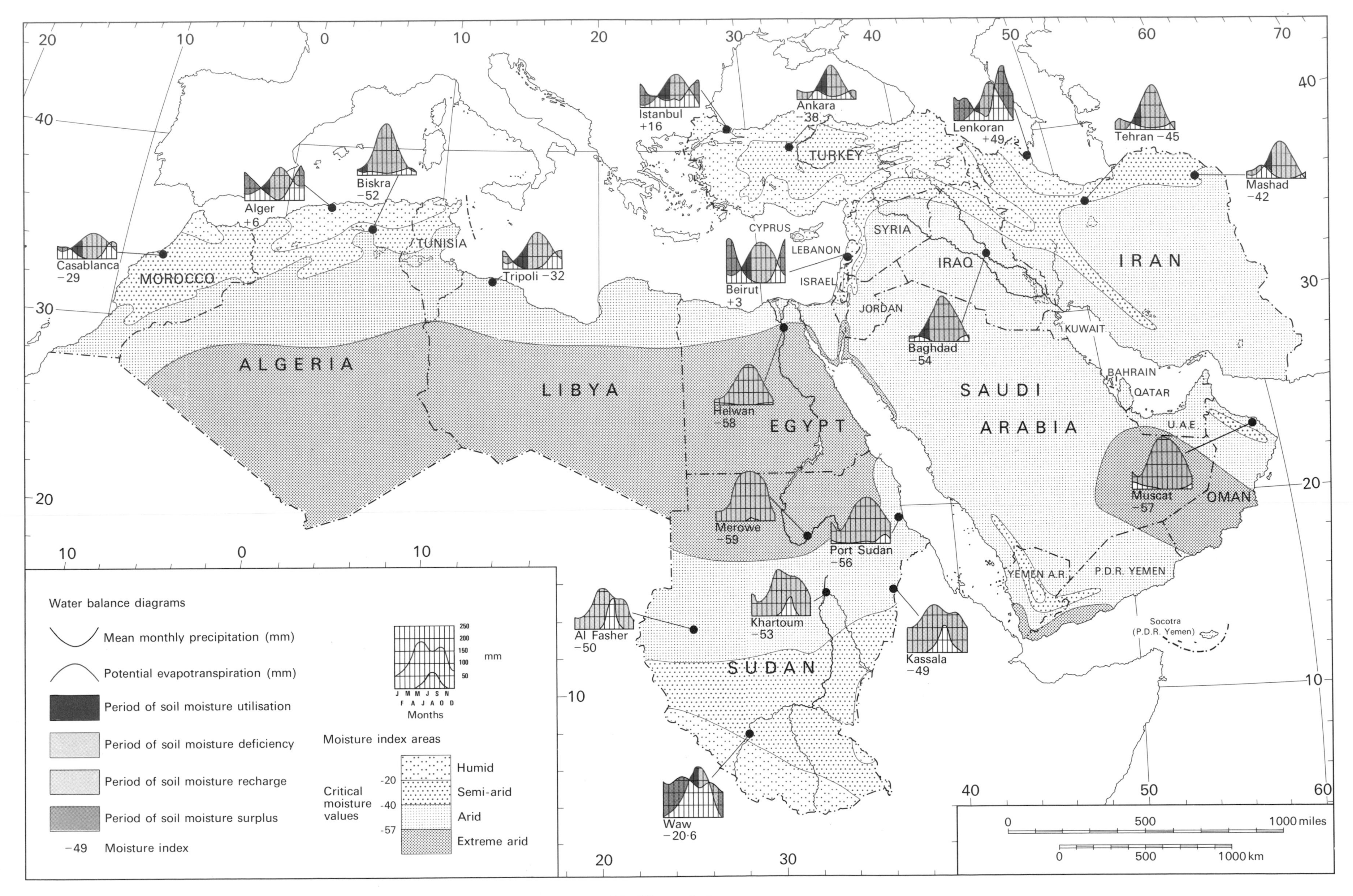

10 Water balance

Water balance

John Dewdney

With the exception of the higher mountain districts, there are few areas in which temperatures are a major constraint on agricultural production; by far the most important environmental factor is the availability of water in terms of both annual total quantity and its seasonal distribution. These depend not only on the annual and seasonal quantities of precipitation but also on its effectiveness, which depends on temperature and the resultant evapotranspiration rates.

The concept of evapotranspiration rates as defined by Thornethwaite is a hypothetical one, defined as the maximum amount of evaporation and transpiration which would occur from a grass-covered surface if an abundant and continuous supply of moisture were available in the upper layers of the soil, rather than an expression of what actually occurs under existing conditions of soil, vegetation and rainfall. Nevertheless, empirical formulae have been developed for the calculation of potential evapotranspiration rates under given temperature conditions and from these have been developed the data on water balance, both seasonal and annual, displayed in Map 10.

The situation at each of 19 stations is recorded by a water balance diagram based on graphs of the monthly values of precipitation and potential evapotranspiration. The chart for Istanbul, where there is year-round rainfall with a pronounced summer maximum, illustrates the four situations which can occur. Precipitation exceeds potential evaporation from January to April and again from October to December. December–April are months of 'soil moisture surplus', during which there is sufficient water available to support the growth of many cultivated plants. In May and June, evapotranspiration exceeds precipitation, but plants can draw on moisture stored in the soil – the period of 'soil moisture utilisation' – and further crop growth may be possible. By July, however, soil moisture is exhausted and potential evapotranspiration is far greater than precipitation; consequently July, August and September are months of 'soil moisture deficiency', when further plant growth can occur only with the aid of irrigation. In October and November, precipitation again exceeds potential evapotranspiration, but in these months the excess is used up in 'soil moisture recharge' and it is not until December that a soil moisture surplus is again available to support plant growth.

By comparing the actual quantities of moisture surplus and moisture deficiency it is possible to compute a moisture index for the year as a whole, and critical values of this index have been used to distinguish the 'humid', 'semi-arid', 'arid' and 'extreme arid' zones indicated on the map by shading.

Humid areas are relatively limited in extent, being confined to parts of the northern Maghreb, the coastal and mountain periphery of Turkey, the Caspian coastal zone of northern Iran, northern Israel, Lebanon and western Syria (i.e. the Levant coast), the extreme south of Sudan and the highest mountain districts of Yemen A.R. In these areas, cultivation without the aid of irrigation is possible over much of the year, but even in this zone there may be periods of intense drought, indicated by large soil moisture deficiencies in certain months. This is particularly true of districts where the climate is clearly Mediterranean, with rainfall confined to the winter months and a summer period during which there is virtually no precipitation. This situation is most clearly exemplified by the water balance diagram for Beirut, where there are large surpluses from December to March and large deficiencies from June to October. Alger shows similar features in a less intense form.

Semi-arid areas lie inland from and/or to the south of the humid zones, in districts where the annual precipitation is considerably less, and cover the southern Maghreb, much of the interior of Turkey, parts of western and northern Iran, a narrow strip of territory inland from the Levant humid zone, and south-central Sudan. The water balance diagram for Ankara is characteristic of the semi-arid zone. The period of soil moisture deficiency extends from July to October (though the scale of the deficiency may be less than at some stations in the humid zone, e.g. Beirut). Because of the relatively modest precipitation, soil moisture recharge takes from November to March. April–June is the period of soil moisture utilisation and there is no significant period of soil moisture surplus. Under these conditions, non-irrigated cultivation may be possible during the winter months, but this has the effect of delaying soil moisture recharge in autumn and speeding up soil moisture utilisation in spring and early summer. Thus irrigation may be necessary to supplement the natural precipitation. A further complication in both humid and semi-arid zones is the occurrence of much of the winter precipitation in the form of snow; large quantities of water are stored as snow in the uplands and released in the spring and early summer when the snow melts. Under these circumstances, the seasonal availability of water may be significantly different from that indicated by the water balance diagrams.

The arid and extreme arid zones cover by far the greater part of the Middle East and North Africa, extending in a broad belt from the Atlantic to the Red Sea and thence across the Arabian peninsula into eastern Iran. Several countries – Libya, Egypt, Saudi Arabia, the Gulf states – are wholly arid, and arid conditions cover the bulk of the territories of several more – Algeria, Tunisia, Jordan, Iraq, Iran, the Yemens and Oman. Turkey, Lebanon and Cyprus are the only countries with no arid area. By definition, the arid and extreme arid zones are characterised by a large annual water deficit. In the more favoured parts of the arid zone there may be a single month with a water surplus, as at Tripoli, and there may be brief periods of soil moisture recharge and soil moisture utilisation, as at Baghdad. But the majority of stations in these zones, particularly those in the extreme arid section, such as Helwan and Merowe, show very low precipitation, very high evapotranspiration rates and a large moisture deficiency in every month. In these areas, cultivation without the aid of irrigation is virtually impossible.

The above discussion of the water balance is based on average conditions of precipitation and temperature across the region. It must be realised that conditions in individual years and seasons may differ considerably from these averages. Monthly temperatures, in most years, do not diverge a great deal from the mean, but rainfall throughout much of the region is characterised by great annual variability, this variability tending to increase as total precipitation diminishes. At Rize, for example, on the Black Sea coast of Turkey and thus in the humid zone, annual precipitation averages 2440 mm, the maximum annual total recorded is 4045 mm and the minimum 1758 mm, a ratio of 2.3 to 1. At Konya, in the semi-arid zone, the mean is 316 mm, with extremes of 501 mm and 144 mm, a ratio of 3.5 to 1. At Baghdad, in the arid zone, the equivalent figures are mean 151 mm, maximum 336 mm, minimum 72 mm, ratio 4.6 to 1; and at Cairo, in the extreme arid zone, they are mean 22 mm, maximum 64 mm, minimum 1.5 mm, ratio 42 to 1.

In addition, it must be remembered that individual crops have their own specific water requirements which are different from, and usually greater than, those of the grass cover envisaged in Thornethwaite's formulation of the water balance. Thus the existence of moisture in the soil at a particular time, as indicated by the water balance diagram, is no guarantee that a particular crop can be grown at that time. The most significant element in the water balance is the period of soil moisture deficiency, the period when no crop can be grown without the aid of irrigation.

KEY REFERENCES

Beaumont, P., 'Water resources and their management in the Middle East' in J. I. Clarke and H. Bowen-Jones (eds.), *Change and Development in the Middle East* (Methuen, London, 1981).

Beaumont, P., Blake, G. H. and Wagstaff, J. M., *The Middle East: A Geographical Study* (Wiley, Chichester, 1976), ch. 2, 'Climate and water resources'.

Thornethwaite, C. W., Mather, J. R. and Carter, D. B., *Three Water Balance Maps of Southwest Asia*, Pubs. Clim. Drexel Inst. Technol., XI (1958).

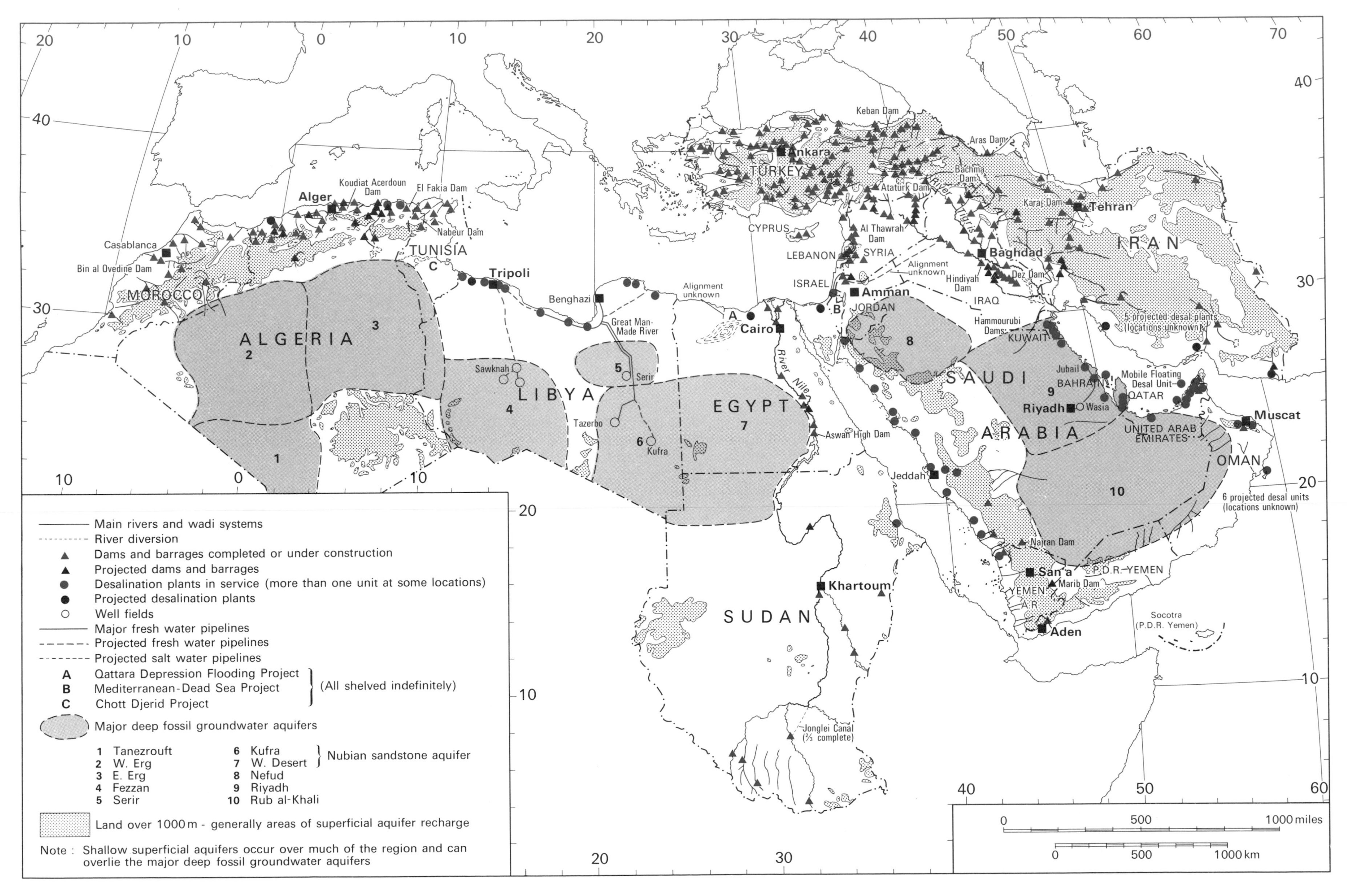

11 Fresh water supply

Fresh water

Jonathan Mitchell

Fresh water supply in the Middle East and North Africa is increasingly critical. An already scarce resource is being stretched by rapidly growing populations and by expanding agriculture and industry. The need to make full use of the naturally available fresh water and to find new sources of supply has never been more acute. This is reflected in the high levels of investment right across the region in everything from dam construction to experimental solar-powered desalters, and even in proposals to modify local climates.

Table 3 shows total water availability (surface and groundwater) within the region on a per capita basis. Taking 1000 m^3 per annum per capita as a minimum ideal availability, it can be seen that all but six countries will face serious water shortages by the end of the century, and even in these water will be far from plentiful. Failure to increase the region's water supply could lead not only to hardship for the people, but also to conflict, especially over international rivers.

Dam and barrage building is the most usual way to store and control surface water. The majority of the region's perennial rivers have now been dammed at least once, and multiple damming is becoming the norm.

The larger dams are generally multi-purpose, combining water storage with hydropower generation and flood control. The largest example of a multi-purpose dam in the region is the Atatürk Dam on the Turkish Euphrates. When completed around 1991 it will have a 2400 MW generating capacity, and will eventually provide irrigation water for 727,000 ha of land. Many of the recent smaller dams are also multi-purpose, but the older ones tend to be just for river control and water storage.

With the exception of the Nile, the perennial rivers are found north of 30°N, and even here numerous water courses are seasonal. Increasingly, seasonal water courses are being dammed for flood control, water storage, and even aquifer recharge. Saudi Arabia has a programme to build 60 dams of various sizes, all on seasonal water courses, by 1990. The other Arabian peninsula states, with the exception of Kuwait, Bahrain and Qatar, have similar if smaller programmes.

The use of groundwater as a fresh water source has been practised for millennia, using springs and wells and, in the eastern Gulf and Iran, subterranean canal systems known as Qanats or Falaj. However, in recent decades the use of groundwater has markedly increased. Many states rely heavily on groundwater, for example, the Gulf states, Libya and Israel. Elsewhere, groundwater is extensively used, especially for irrigation, even though the overall reliance may be less. But groundwater as a source of fresh water is not without its problems.

The most serious of these is the question of aquifer recharge. Overpumping and exploitation of shallow aquifers, particularly near the coast, can lead to salt water ingress and aquifer contamination. This is getting quite serious in Bahrain and Qatar, for example. In the Nile delta the problem is exacerbated by reduced water levels in the Nile preventing normal recharge. Salt water has now reached as far as 20 km inland. Schemes for aquifer recharge are under investigation in several countries: for example, dams on porous rock for this purpose are being tried in Oman. Excessive pumping related to recharge also increases the cost of extraction as the water table falls.

The deep fossil water aquifers present their own problems. At least one school of thought sees these as non-renewable resources. This suggests that the huge Libyan investment in the Great Man-Made River project is only a medium-term solution to that country's fresh water problems. Added to this, the water from the deep aquifers is frequently brackish. This means that it either has to be desalted (hence inland desalination plants) or mixed with desalted sea water to form a palatable cocktail (in theory), as is the case with Riyadh's fresh water.

Table 3. *Per capita water availability (000m^3/annum)*

	1971	2000	% population increase
Algeria	2.2	1.0	111
Arabian peninsula	0.7	0.3	106
Cyprus	0.06	0.05	22
Egypt	0.1	0.05	111
Iran	6.0	2.5	145
Iraq	3.6	1.3	173
Libya	3.7	1.2	198
Morocco	2.1	1.9	132
Sudan	4.0	1.9	107
Syria	3.0	1.0	165
Tunisia	0.9	0.4	126
Turkey	4.9	9.3	118

Source: U.S. Council on Environmental Quality, *Global 2000 Report to the President* (Penguin Books, Harmondsworth, 1982).

There is clearly a need for other sources of fresh water. Desalination has been seen as something of a panacea in this respect, but with its high costs and the current state of desalting technology it is not. It is at present prohibitively expensive for large-scale agriculture, and is used primarily as a potable source. Desalted water costs at least $2 per m^3 when produced by conventional means, though solar pond techniques under development in Qatar could cut this by two-thirds. Despite the cost the region had, in 1982, 65% of total world desalting capacity, and 35% of the world's desalination plants.

The use of treated sewage for garden irrigation, and the irrigation of crops which require cooking before consumption, is being tried in Jordan and Kuwait. Considerable caution has to be exercised to avoid the transmission of diseases. Despite this, Kuwait has plans to irrigate 16,000 ha in this way, thereby releasing valuable desalted water for other purposes. The use of this source of water is set to expand, especially as the Islamic clergy have stated that there is no religious objection to its use.

Other solutions to the problem of fresh water supply have been many and varied. Towing icebergs from the Antarctic is technically feasible but might well prove as expensive as desalted water. Bringing water to the region by tanker from Europe and the Philippines has been suggested and, while it may happen, heavy reliance on an outside supplier for something as vital as water would be politically undesirable.

It is clear that there is no one solution to fresh water supply shortages in states where surface water and naturally recharged groundwater do not meet demand. Careful management of available sources and conservation practices throughout the region should help prevent severe shortages. This, combined with a diversification of sources and investment in new technology, particularly in the field of desalination, could make the future of fresh water supply secure.

KEY REFERENCES

Charnock, A. *et al.*, 'Water resources', business feature, *M.E.E.D.*, 10 Aug. 1984, pp. 27–43.

Gischler, C., *Water Resources in the Arab Middle East and North Africa* (Menas Press, London, 1979).

Keen, M., 'Cheaper, purer water from the sun', *Middle East Water and Sewage*, Sept./Oct. 1985, pp. S14–S16.

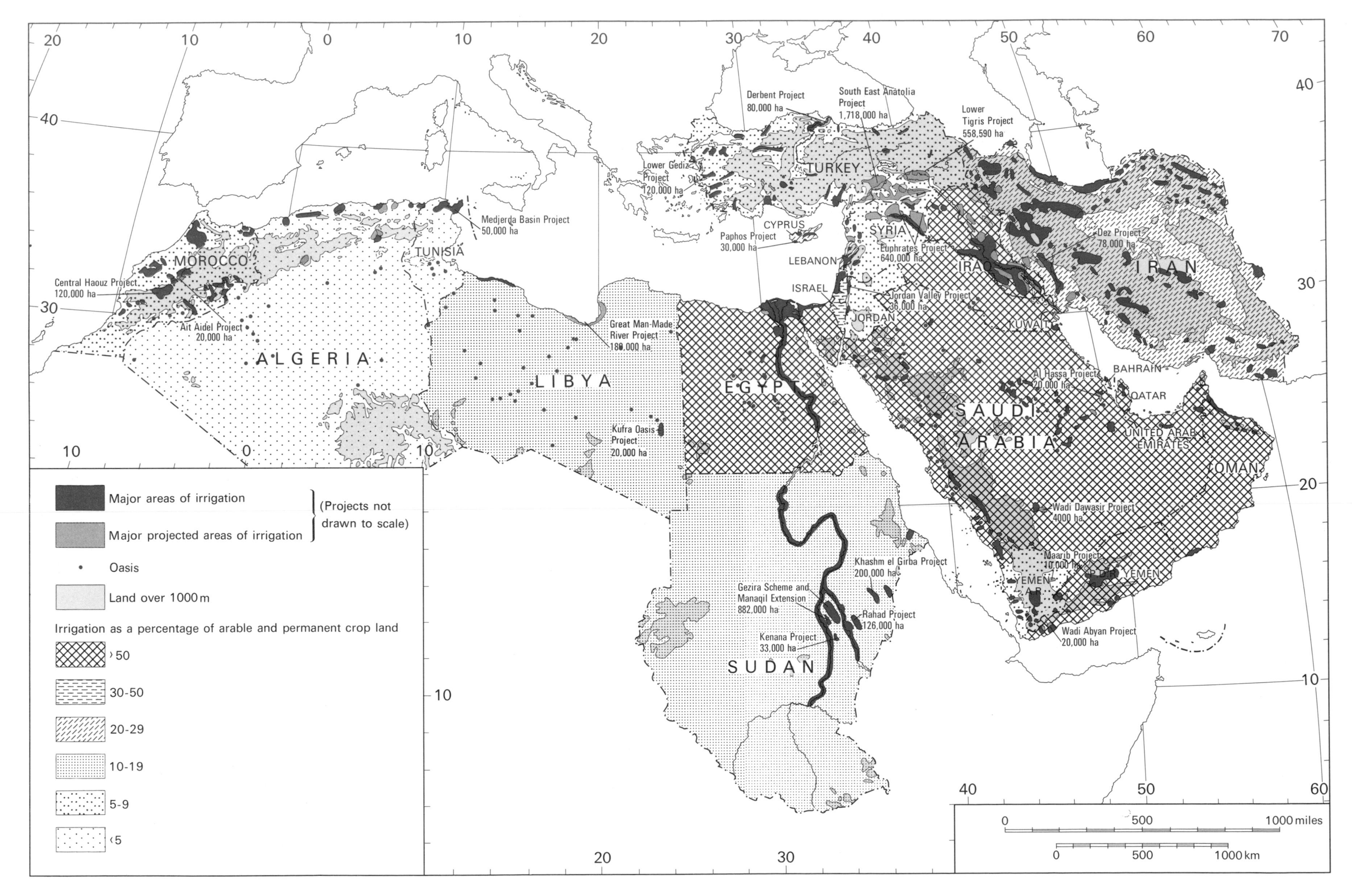

12 Irrigation

Irrigation

Jonathan Mitchell

The practice of irrigating crops is at least as old as the earliest civilisations. Major irrigation schemes were begun in the latter half of the fourth millennium B.C. in the Tigris–Euphrates lowlands, and on the floodplain of the lower Nile, but there is some evidence to suggest that a form of irrigation may have existed at Jericho around five and a half thousand years before this.

Traditionally, irrigation has been practised in relatively restricted areas, beside rivers and around oases and springs; indeed wherever fresh water was plentiful and could be easily diverted on to the land. Aqueducts, canals and tunnels have all been used to carry irrigation water since antiquity, but the scale of water usage has in the past never been very great.

In recent decades the amount of land under irrigation throughout the region has increased dramatically as a direct response to ever-increasing demands for agricultural products. This demand is partly a reflection of population increases, but factors such as the desirability of agricultural self-sufficiency and the need for exportable products all play a part. While not the only element in agricultural expansion, irrigation is by far the most important, being vital to increase yields (vertical expansion) and to increase the cropped area (horizontal expansion).

Much of the region is arid or semi-arid, and dry farming is largely confined to those areas with over 300 mm of rainfall per annum. For example, the basic food staple, wheat, will only grow successfully in areas with over 320 mm of rainfall, and ideally requires 450–650 mm if it is to be successful commercially.

Many of the major irrigation projects lie outside the 300 mm isohyet and are clearly for horizontal agricultural expansion. One of the largest single projects in this area is the 640,000 ha Euphrates project in Syria. This is considered vital to the long-term viability of the Syrian economy.

Recent fresh water supply technology (Map 11) has made the opening up to agriculture of desert and semi-desert areas possible. Kufra Oasis in Libya is a 20,000 ha project using groundwater for irrigation in the Sahara, and the Libyans currently have in hand a further 180,000 ha irrigation project using Saharan groundwater. Much of the new irrigation in the Arabian peninsula also uses groundwater, though this is a finite resource, and requires careful management. To a lesser extent desalted water is also used, but this is generally prohibitively expensive, given current technology, for all but the most intensive agricultural projects. To make the fullest possible use of irrigation water in areas of extreme scarcity and high potential evaporation, new water application techniques have been and are being developed. Of these perhaps the best known example is hydroponics.

Vertical agricultural expansion is equally significant. The largest new irrigation project in the region, if not the world, the South East Anatolia project in Turkey (about 1.7 m ha), is concerned with this, as are practically all irrigation projects in areas with over 300 mm of rainfall. It has been demonstrated in Syria that wheat will yield up to 86% more under irrigated conditions than is possible under rain-fed conditions. With barley, the difference is 92%, and with cotton up to a staggering 366%. However, cotton is not generally grown under rain-fed conditions whereas the other two crops are.

There are many problems associated with irrigation, ranging from the cost to technical and physical problems, and political considerations, the latter being most significant where international rivers are being exploited.

The sheer cost of major projects can have serious economic ramifications. The Great Man-made River project in Libya is costing around $500 m a year at present and is adversely affecting other developmental sectors such as industry; however, the long-term benefits should make the investment worthwhile. Poorer countries have to rely on substantial foreign loans to develop their full irrigation potential. Turkey is being assisted by the World Bank for a number of projects, but has experienced difficulty in securing loans for the South East Anatolia project because of the political sensitivity of the potential downstream effects on Syria and Iraq. Similarly, the Syrian Euphrates project has been delayed time and again by political problems, both national and international. The success of irrigation projects in all three major international river basins, the Nile, Tigris and Euphrates, as well as lesser ones such as the Yarmouk and Orontes, all depend heavily on good inter-state relations.

Perhaps the most serious problems are technical and physical. Waterlogging and resultant soil salinity have been serious drawbacks in irrigated agriculture for at least 5000 years. The rise and fall of early Mesopotamian civilisations appears to have been inexorably bound to this. In the late 1970s Syria was losing 300 ha of already irrigated land to soil salinity each year and Turkey was experiencing considerable delays in the full realisation of certain of its major irrigation projects.

To combat this problem a number of measures can be taken. Effective drainage in new project areas and its installation in areas already affected is the best solution. The drilling of tube wells to lower water tables is another method which has been adopted. But for waterlogging to be completely prevented, irrigation practices need to be carefully controlled. Traditional basin irrigation is both wasteful and potentially dangerous. More in favour are sprinkler systems, both fixed and centre pivot, but even these are generally wasteful, especially at times of high evaporation. Individual plant drip feed and hydroponics cut down both waste and salinity risk, but would be too expensive for extensive agriculture.

Waterlogging and high dissolved salt levels are not the only physical problems. Nutrient leaching, especially on desert soils, and the undermining of concrete irrigation canals on gypsiferous soils are also factors which have affected irrigation in the region. On top of this there are numerous problems related to land use and farming practices. It is essential to educate farmers to make sound use of irrigated land.

In conclusion, the region has considerable agricultural potential, much of it as yet unrealised. Irrigation, whether it is in large projects or at individual village level, holds the key to feeding the rapidly growing populations within the Middle East and North Africa. It is not all that is required to increase agricultural production, but it is arguably the most important factor.

KEY REFERENCES

Charnock, A. *et al.*, 'Water resources', business feature, *M.E.E.D.*, 10 Aug. 1984, pp. 27–43.

Keen, M., 'Cheaper, purer water from the sun', *Middle East Water and Sewage*, Sept./Oct. 1985, pp. S15–S16.

O'Sullivan, E., 'Water decade goals recede as population grows', *M.E.E.D.*, 26 March 1982, pp. 16–21.

Ritchie, M., 'Libya: taking the plunge with the G.M.R.', *M.E.E.D.*, 20 July 1985, pp. 14–16.

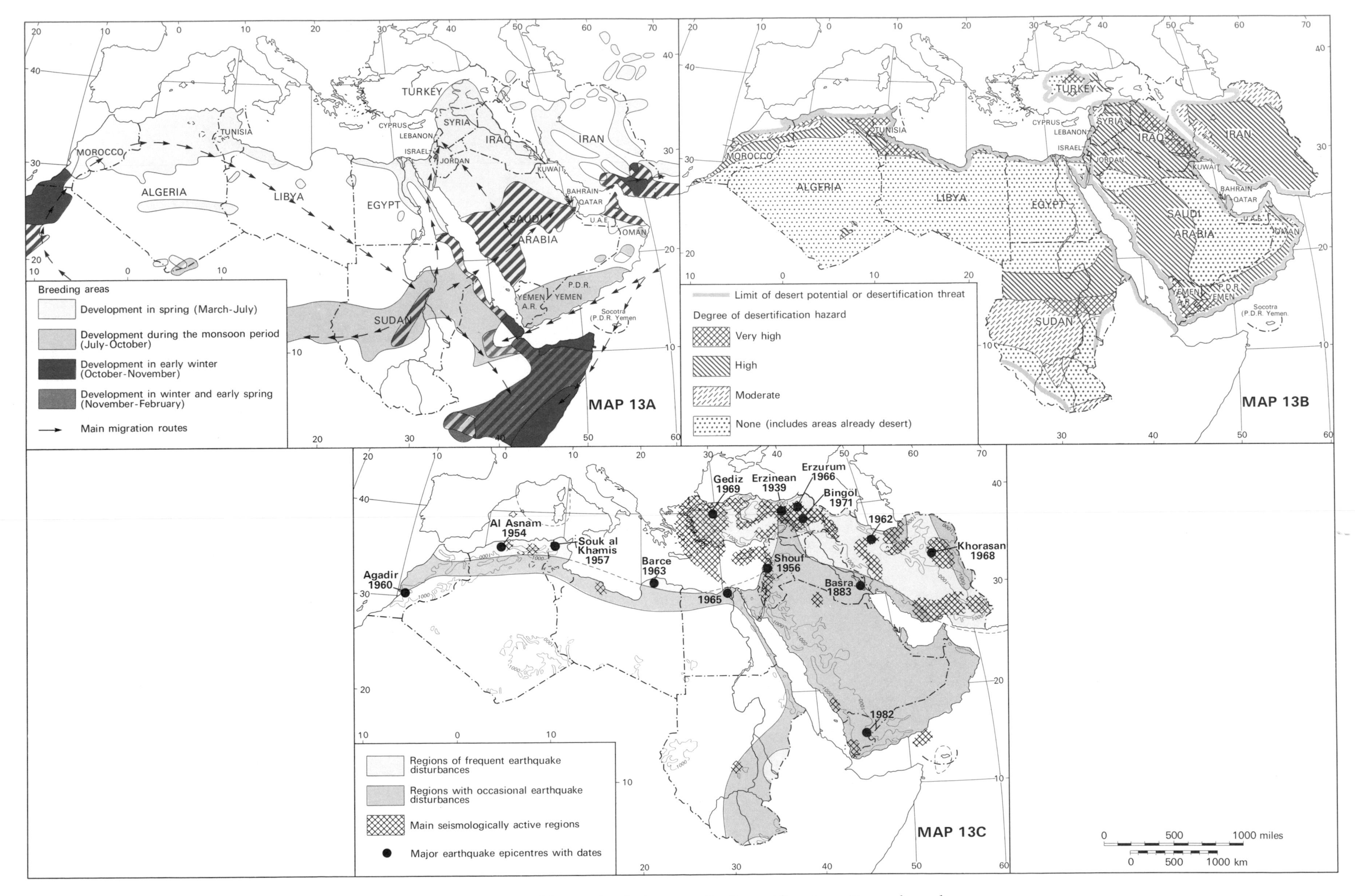

13 Environmental hazards: A. locusts; B. desertification; C. earthquakes

Environmental hazards

John Dewdney

The Middle East and North Africa face several major environmental hazards, three of which are illustrated in Map 13.

Earthquakes

In terms of loss of life, injury and damage to property, by far the most serious hazard, and the most unpredictable, is that presented by earthquakes. At least 160,000 people have been killed by earthquakes in the region during the twentieth century, including some 60,000 since 1960.

The region is one of those areas of the world in which seismic activity is most frequent and most intense; why this should be so is indicated by Fig. 2 above. The basic structural arrangement is one in which relatively stable blocks – the European to the north and the Afro-Arabian to the south – lie on either side of the belt of orogenic activity where the sediments of the ancient Tethys Sea have been compressed, by the movement of the blocks, into a complex system of fold mountains. The differential movement of the two blocks continues and it is in the intervening zone that seismic activity has been most intense and where the most devastating earthquakes of recent decades have occurred.

There have been more than a hundred earthquakes measuring 5 or more on the Richter scale since 1950. The effects of individual earthquakes are not directly related to the magnitude of the event as measured by the Richter scale; the actual occurrence of death, injury or damage to property is largely a matter of chance, connected with such factors as the location of the epicentre (whether in a densely populated or empty area), the time of day (determining whether the people are indoors or not) and, above all, the capacity of buildings to withstand the shock. Thus, in more than half the recorded earthquakes, loss of life or injury has occurred on a very minor scale or not at all and damage has been very limited.

Of all the countries of the region, the most seriously affected has been Iran, which has experienced more than 50 earthquakes since 1950, causing the deaths of some 35,000 people. The most severe occurred in the Elburz Mountains in 1957 and 1962 and near Khorasan in 1968. Turkey has experienced a similar number though loss of life (*c.* 7000) has been smaller; the worst were in Erzurum province in 1959, 1966 and 1967 and Bingöl in 1971. Earthquakes have been less frequent in the Maghreb but the loss of life has exceeded 20,000; some 15,000 were killed in the Agadir earthquake in 1960 – the worst anywhere in the region for many decades.

Seismic activity also occurs along the faults which separate the African and Arabian blocks. In Lebanon in 1956 136 were killed and over 2500 in Yemen in 1982. Movement along quite minor faults in Libya in 1963 led to 300 deaths.

Almost any part of the region is likely to experience a significant earthquake within the next 20–30 years. Systematic collecting of information via a more widespread network of seismological stations is necessary to estimate the likelihood of an earthquake occurring in a particular area; special – and expensive – building techniques are necessary to reduce the effects of future shocks when they come.

Locusts

Swarms of locusts occasionally descend upon cultivated areas and devour the crops wholesale. A swarm may cover many square kilometres and each square kilometre can contain as many as 50 million locusts, which can consume as much as 100 tons of vegetation in a day.

The desert locust, which is the predominant species in the region, is endemic to large areas, being absent (despite its name) only from the drier parts of the Sahara and Arabia and from the wetter parts of Turkey, Iran and Sudan. In order to breed successfully, the female locust must lay her eggs (usually numbering 20–100) in damp ground containing a soil moisture reserve equivalent to about 20 mm of rainfall. This moisture permits the egg to hatch to the nymph or 'hopper' stage and supports the insect over several weeks during which it develops to the adult winged phase. As the map shows, this development occurs at different times of the year in different parts of the region according to the local rainfall regime.

Once mature and able to fly, the locust may live in two different forms or 'phases'. In the 'solitary phase' it lives alone or in small groups and is indistinguishable in its behaviour from other species of grasshopper. When numbers increase above a certain level however, the locust may enter the 'gregarious phase', moving about in huge masses or swarms, and it is these which create a major hazard to crops and thus to food supplies.

Swarms (or 'plagues') of locusts in the gregarious phase can fly very long distances, covering 100 km or more a day and consuming the vegetation of several square kilometres in a few hours. The main directions of migration are shown on the map. The most important routes within the region are the circulatory movement from Sudan to the Atlantic coast (flying to the south of the Sahara) and thence across North Africa back to the Sudan; and from Sudan and the Arabian peninsula northwards into the Levant and Mesopotamia. Swarms from the Indian subcontinent sometimes reach southern Arabia and the Horn of Africa. The general direction of these migrations is influenced by the prevailing winds, and the speed of movement by wind strength and reliability.

The most promising method for the prevention of locust plagues is to attack the locusts with pesticides in their breeding areas so that numbers are kept down below the level at which they are likely to enter the gregarious phase.

Desertification

Desertification has been defined as 'the spread of desert-like conditions in arid or semi-arid areas with less than 600 mm of precipitation due to man's influence or to climatic change'. Thus desertification is a potential hazard throughout much of the region. Apart from areas already desert, the threat is absent only from coastal sections of the Maghreb and Levant, the wetter parts of Turkey and Iran, southern Sudan and a few small upland areas in the Atlas, Cyrenaica and Yemen. The map distinguishes degrees of hazard and it will be observed that the great bulk of the areas where the hazard exists are allocated to the 'high' or 'very high' category. The degree of hazard is dependent largely on the soil, vegetation and water balance conditions but is also influenced by human action.

There are conflicting opinions on the role of climatic change in the desertification process. While some argue that the region is now in a phase of a long-term increase in aridity, others see no long-term trend but merely a continuation of a sequence of fluctuations involving series of abnormally dry and abnormally wet periods. Even if the latter view is correct, the desertification which occurs during dry periods is seldom reversed during the wet, and thus the desert expands intermittently.

Recent decades have witnessed rapid population growth throughout the region. This has resulted in increases in livestock populations, in the demand for crop products and in the use of the natural vegetation as fuel. Overgrazing, expansion of cropping into marginal areas and cutting of the vegetation result in a deterioration of the soil and vegetation cover, soil erosion and other physical processes which lead to desertification. Desertification is due primarily to the increasing pressure of man and his animals on fragile and unstable ecosystems.

KEY REFERENCES

Heathcote, R. L., *Arid Lands, Their Use and Abuse* (Longmans, London, 1983).

Pedgley, D. E. and Symmons, P. M., 'Weather and the locust upsurge', *Weather*, 23, 1968, pp. 484–92.

Perera, J., 'On the eve of destruction', *Middle East*, 105, 1983, pp. 25–33.

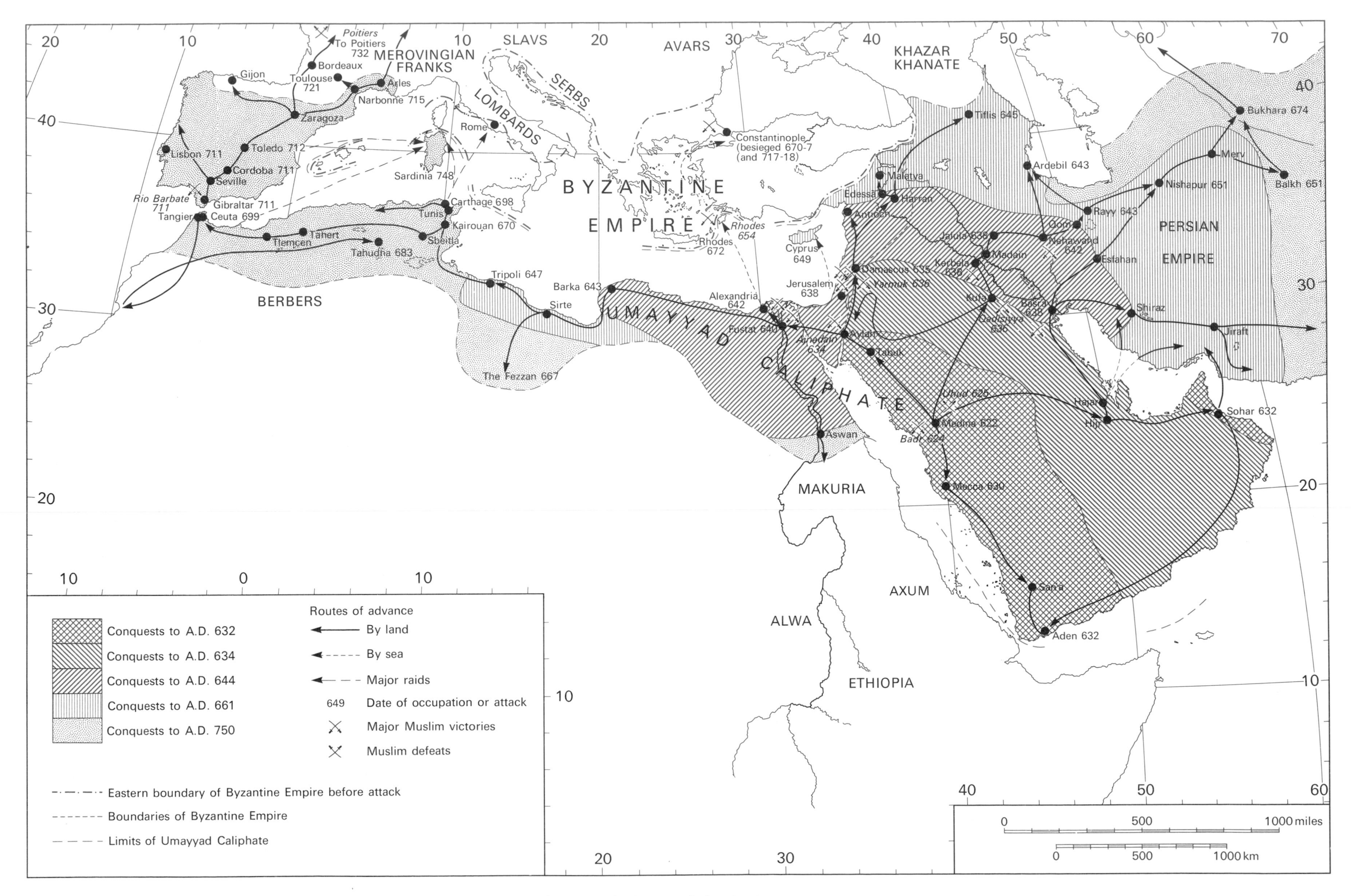

14 The rise of Islam and the Arab conquests

The rise of Islam and the Arab conquests

Richard Lawless

In the late fourth century A.D. the eastern and western halves of the Roman Empire became separate political entities and were never again fully reunited. The Western Empire fell to the Germanic invaders in the fifth century, and the last puppet emperor in the West was deposed by the barbarians in A.D. 476. The Eastern or Byzantine Empire, however, with its capital at Constantinople (New Rome, founded in A.D. 330), survived the barbarian onslaught and developed its own distinctive identity. The Byzantine state, consisting of the provinces of Asia Minor, Syria and Egypt, was more highly developed, more populous and more prosperous than the western provinces. Christianity, which had been adopted as the official religion in A.D. 313, was deeply rooted. Unfortunately, an attempt to recover the western provinces during the sixth century greatly weakened the empire, and in the early seventh century Sassanid forces penetrated deep into Byzantine territory. Both empires emerged exhausted. Years of warfare had resulted in much material destruction and crippling taxation of the native peoples. The old empires were in no state to resist an event which changed the whole course of history in the middle East: the Arab conquest.

The Arabs of Arabia were by no means isolated from the civilised world of the Byzantine and Persian Empires. They had come into contact not only with material goods from outside but also with Christianity and Judaism. Primitive Islam was carried to all parts of the Middle East by nomads, but the religion was born in the towns of western Arabia where the tribes who had settled in these centres needed to evolve a new value system more closely adapted to their new way of life. Muhammed was born around A.D. 570 into a community of merchants, the Quraysh tribe, who in the sixth century had gained control of the trading city of Mecca. He was thus a citizen, not a nomad. The call came to him in his fortieth year. He preached submission (Islam) to the will of one God, Allah, of whom Muhammed was the chosen prophet. Only slowly did he win support for his religious claims, and in A.D. 622 he was actually driven from Mecca and forced to take refuge in Medina. This flight (*hegira*) from Mecca to Medina is taken as the beginning of the Islamic era.

Mecca was eventually retaken, and Islam was able to unify the Arab tribes in the Arabian peninsula. After Muhammed's death in A.D. 632, within 80 years they conquered an area far larger than that amassed by the Romans in 800 years. But the Arab conquests arose in an unplanned way and were not part of a systematic campaign. They appear to have begun rather thoughtlessly as a reaction to local events and then to have broadened out. This rapid thrusting out of the Arabs from Arabia had little to do with specifically religious motives such as the conversion of the region to Islam, but the real causes remain obscure. What is certain is that to the oppressed subject populations of the Byzantine and Persian Empires the Arabs appeared as liberators, a key factor in the rapid success of small Arab armies with no overall strategy against superior forces of the Byzantines and Sassanids.

The first phase of Muslim expansion took place during the decade following the death of Muhammed. The conquest of Mesopotamia, later known as Iraq, began in 633. The Persians resisted the Arab advance across the Euphrates but were defeated at the great battle of Qadisiyya in 636. By 642 the Arabs were masters of Iraq and all of western and central Persia. Palestine was occupied in 634 after the battle of Ajnadain and early in 635 Damascus was besieged, and captured in 636. In 639 northern Syria passed under Muslim control and by 640 the occupation of Syria was complete. The conquest of Egypt began in 639. Heliopolis was occupied in 640 and Alexandria in 642. The Arabs conquered Upper Egypt and even penetrated into Cyrenaica. The Arab advance only halted when it reached major natural obstacles: the Taurus Mountains in the north, the arid expanses of central Iran and the deserts of Cyrenaica in the west.

The second phase of expansion occurred during the Umayyad Caliphate when the Arabs reached the Atlantic in the west, advanced into central Asia, and in the east penetrated as far as the Indus river. Arab expeditions were sent in three directions, to Constantinople and Asia Minor, Central Asia, and into North Africa and Spain. The Arabs did not seek to occupy Asia Minor but were content to carry out seasonal raids into the territory. The prestigious Byzantine capital, Constantinople, however, exerted a powerful attraction for the Arabs but, although the great city was twice besieged by Muslim forces in 670–7 and 717–18, it did not fall to the invaders. The province of Khorasan in northeast Iran served as the springboard for Arab advances into Central Asia and towards India. Afghanistan was conquered in 699–700 and successful expeditions were mounted to Tokharistan (705), Sogdiane (706–9), Khwarezm (710–12) and Ferghana (713–14). Bukhara and Samarkand became the great Muslim cities of Central Asia. To the south, Baluchistan was conquered in 710, Sind in 711–12, bringing the Arabs to the banks of the Indus, and the southern Punjab in 713. Arab expeditions had been mounted into northwest Africa in 643 and 647, and in 670 a permanent military camp was established at Kairouan. But Ifriqiya and Tripolitania were abandoned in 683 and fifteen years were to elapse before a new expedition resulted in the capture of Carthage in 698. Between 705 and 708 Arab armies swept across the central and western Maghreb to the Atlantic and on into Spain in 711. Continuing northwards they were only finally repulsed at Poitiers in 732. The Arab tide had turned but the conquerors retained control of much of Spain. Indeed it was in Al-Andalous or Muslim Spain that the Arab imperial system yielded some of its most remarkable results. Al-Andalous with its magnificent capital at Cordoba, one of the great Muslim cities, became the most civilised area of Western Europe.

At the outset, what the Arabs did with their sword was not to spread Islam but merely their own political rule, and there was a gap of over a century between the political conquest of the Middle East by the Arabs and the beginnings of the Islamisation of the region. The Arab conquerors had no interest either in the religious conversion of their new subjects or in mingling with them. The non-Muslim subjects bore the main burden of taxation, and in return the Arabs made no attempt to interfere with their local customs, civil or religious. To the Christian populations of Egypt and Syria the tolerance of the new Muslim rulers was on the whole preferable to the fanatical oppressiveness of the Byzantines. As the centuries progressed, however, the number of converts to Islam began to multiply, and the Arab aristocracy were gradually merged with other ethnic groups who, though non-Arab, were nevertheless Muslim. The two categories of Arab and non-Arab were replaced by the broader notion of being a Muslim. The Islamisation and Arabisation of the population of the Middle East was reflected in the collapse of the Arab regime of the Umayyads, who had held the Caliphate since the early seventh century, and its replacement in the mid-eighth century by the cosmopolitan empire of the Abbasids in which non-Arab and especially Persian influences were dominant. With the succession of the Abbasid dynasty in A.D. 750 the centre of Muslim power moved eastwards from Damascus, the Umayyad capital, to Baghdad, and there was a new orientation away from the Mediterranean towards the advanced civilisations of China and India.

KEY REFERENCES

Donner, F. M., *The Early Arab Conquests* (Princeton University Press, Princeton, 1981).

Gabrieli, F., *Muhammad and the Conquests of Islam* (Weidenfeld and Nicolson, London, 1968).

Mantran, Robert, *L'Expansion musulmane VIIe–XIe siècles* (Presses Universitaires de France, Paris, 1969).

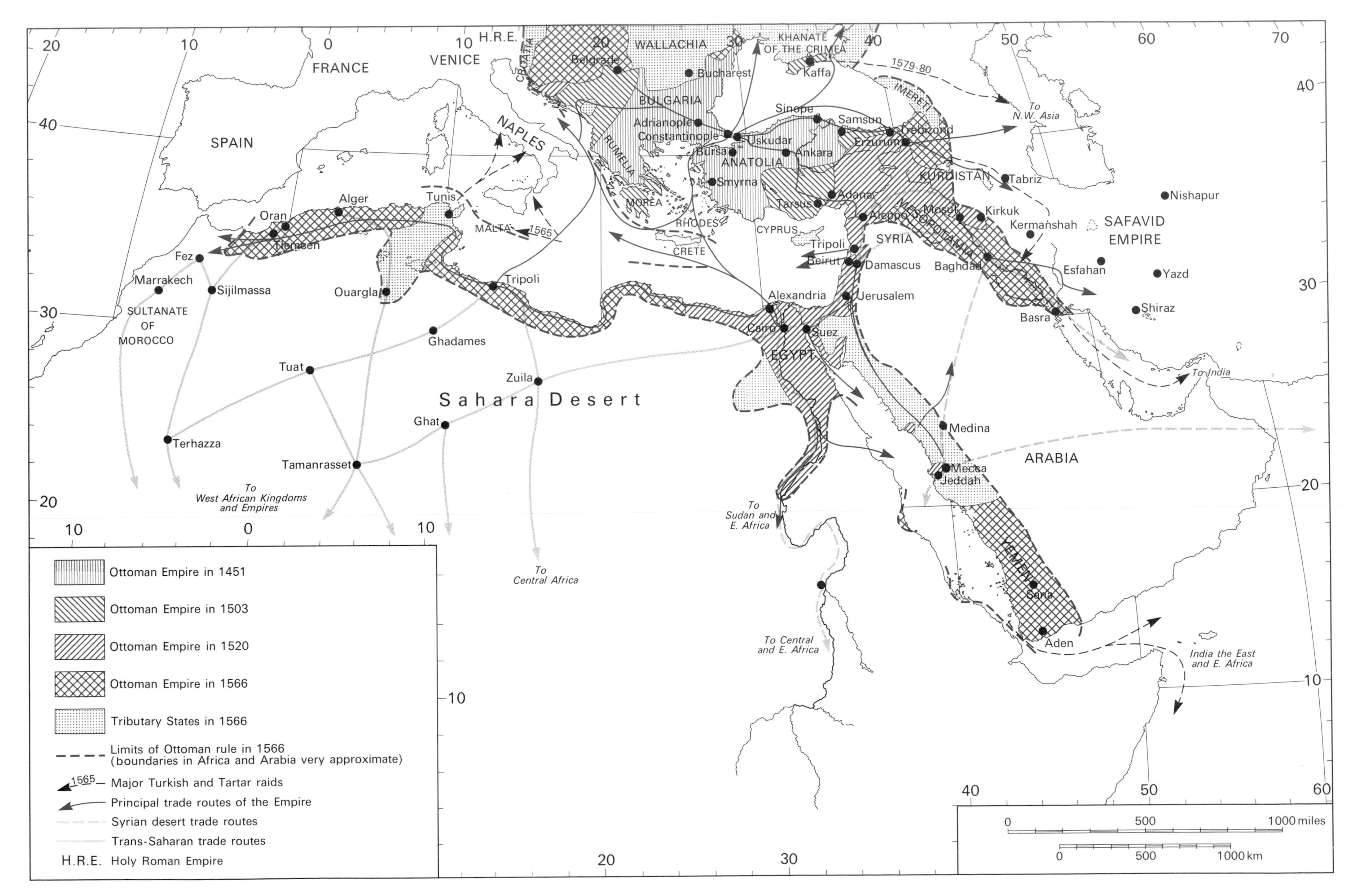

15 The Ottoman Empire to 1566

The Ottoman Empire

Richard Lawless

The weakening of centralised political power led the Abbasid Caliphs to become increasingly dependent on military power. As the government became more militarised, a new people, the Turks originating in Central Asia, became more and more prominent as a prop of the central regime. First recruited by the Abbasids as military slaves, the Turks rapidly gained control of the caliphs' military forces, and later the whole tribes of Turkish nomads began to infiltrate the empire. By the middle of the tenth century the Abbasid caliphate had lost all real power, and from that time to the end of the nineteenth century various Turkish groups were to establish their political control over most of the Middle East. In the early eleventh century one such nomadic group, the Seljuk Turks, who had embraced orthodox Islam in the previous century, entered Baghdad and established their authority over Iran, Iraq and Syria, and wrested much of Asia Minor from the Byzantines.

The Seljuk Sultanate broke up in the thirteenth century under the impact of the first Mongol invasion, and one of the successor states which emerged was a small Turkish principality founded by the Osmanlis in the northwest of the plateau adjoining land still held by the Byzantine Empire. The new state was able to attract numerous Turkish adventurers to its service and began to expand at the expense of both the Byzantines and its Turkish neighbours. Bursa, an East Roman city, was their first objective and was captured in 1326. In 1353 the Osmanlis (or Ottomans), who had become Islamised and strongly influenced by Persian culture, invaded Europe and in 1361 established their capital at Adrianople (Edirne). By 1400 the Ottomans controlled most of Anatolia while their northern frontier lay along the Danube. But in 1402 they were decisively defeated by a new Mongol invasion led by Timurlane and they lost their possessions in Anatolia. The defeat and dismemberment of 1402 closes the first phase of Ottoman history. The state survived, however, in the Balkans, and in less than a century the Ottomans had recovered the whole of Anatolia, established their control over the northern shores of the Black Sea and, after a siege lasting two months, had taken Constantinople by storm in 1453 – a date considered by many historians to be one of the turning points which mark the end of the middle ages and the beginning of modern times.

Having taken Constantinople the Ottomans were free to direct their ambitious gaze towards prospects of further conquest. They set out after 1455 to round off their Balkan holdings and to complete their control over Asia Minor, while in 1475 the Crimean Khanate became a vassal state under Ottoman suzerainty; thus closing the last remaining route from Western Europe to India and the Far East not under Ottoman or Mamluk control and giving new urgency to the European quest for new routes to India. But to the east Ottoman advances were blocked by the new centralised monarchy of the Safavids, who after 1500 restored Persia to its status as a nation-state. The Safavids became the main enemies of the Ottomans, mainly on religious grounds (the Safavids were Shi'a), and in 1514 at the battle of Chaldiran the Ottomans defeated the Safavid Shah and temporarily occupied his capital, Tabriz. Freed from further danger in the north the Ottomans were in a position to advance against the Mamluks of Egypt and Syria who claimed to be the principal power in the Muslim Near East and looked upon the Ottoman dynasty as upstarts. Damascus fell to Ottoman forces in 1516 and Cairo in the following year. All of Egypt and its dependencies, including the titular guardianship of the Muslim holy places in the Hejaz, passed into Ottoman hands. The Ottoman Sultan assumed the title of Caliph which was retained by his successors until 1924.

When Sulayman, the greatest of all the Ottoman Sultans, came to the throne in 1520 there were three immediate strategic objectives for the consolidation and eventual expansion of the immense territories under Ottoman rule: to secure the northwest frontier against the Habsburgs, to win naval command of the Mediterranean Sea, and to maintain a stable equilibrium with neighbouring Iran. An invasion of Hungary in 1526 brought the greatest part of that country into the Ottoman sphere of influence. Buda was conquered in 1529 but although Vienna was besieged Ottoman forces were forced to withdraw on account of difficulty supplying the troops. In the Mediterranean, Christian and Ottoman naval forces had been at war since 1525. Ottoman sea power had been greatly reinforced by the collaboration of the famous corsair Barbarossa who had given allegiance to the Sultan in 1518. In 1529 he finally drove the Spaniards out of Alger and the so-called 'Barbary states', the regencies of Alger, Tunis and Tripoli, passed under Ottoman suzerainty, making the Ottoman empire incomparably the strongest Mediterranean power. Hostilities with Persia were resumed in 1534 and the Ottomans occupied Baghdad and succeeded in gaining control of Iraq. In 1538 a naval expedition established Ottoman hegemony in the southwestern tip of Arabia. Another war with Persia ended in 1555, leaving Iraq in Ottoman hands and Georgia divided into Ottoman and Persian spheres of influence. Sulayman's death in 1566 is reckoned by modern historians to mark the commencement of the Ottoman Empire's long decline. Its external expansion had not yet come to a halt but its internal condition had begun to show dangerous symptoms of decay.

The Ottoman Sultans made no attempt to create a single unified state. The different races and religions within the empire did not merge into an organic society but remained apart and distinct. In particular, Muslim and non-Muslim subjects were divided on the basis of religion into autonomous communities known as *millets*. Members of each *millet* were free to practice their own faith and to retain their institutions, laws and traditions under the direction of their own religious leaders. In this way the Ottoman ruler was able to exploit the wealth produced by his subjects with the least possible resistance or friction. The Ottoman ruling class – the officials who administered the empire's far-flung territories, and the soldiers (*janissaries*) of the standing army – were all slaves of the Sultan, recruited exclusively at the outset from Christian families and taken as young boys to Istanbul where they were converted to Islam and carefully trained for their future positions. In the seventeenth and eighteenth centuries, however, the Ottoman Empire was in its long period of decline. Military expansion came to an end, and the Sultan was compelled to cede extensive territories in the Balkans and the Crimea after humiliating defeats by more powerful Christian powers, Austria and Russia. Under the impact of these territorial losses the Ottoman system of military organisation, civil administration, taxation and land tenure, which were all geared to the needs of a society expanding by conquering and colonising new lands, broke down and never succeeded in adopting new forms to suit the changed conditions.

KEY REFERENCES

Kinross, Lord, *The Ottoman Centuries: The Rise and Fall of the Turkish Empire* (Cape, London, 1977).

Kissling, H. J. *et al.*, *The Muslim World: A Historical Survey*, Part III: *The Last Great Muslim Empires* (E. J. Brill, Leiden, 1969).

Shaw, Stanford, *History of the Ottoman Empire and Modern Turkey*, vol. 1: *Empire of the Gazis: The Rise and Decline of the Ottoman Empire, 1280–1808* (Cambridge University Press, Cambridge, 1976).

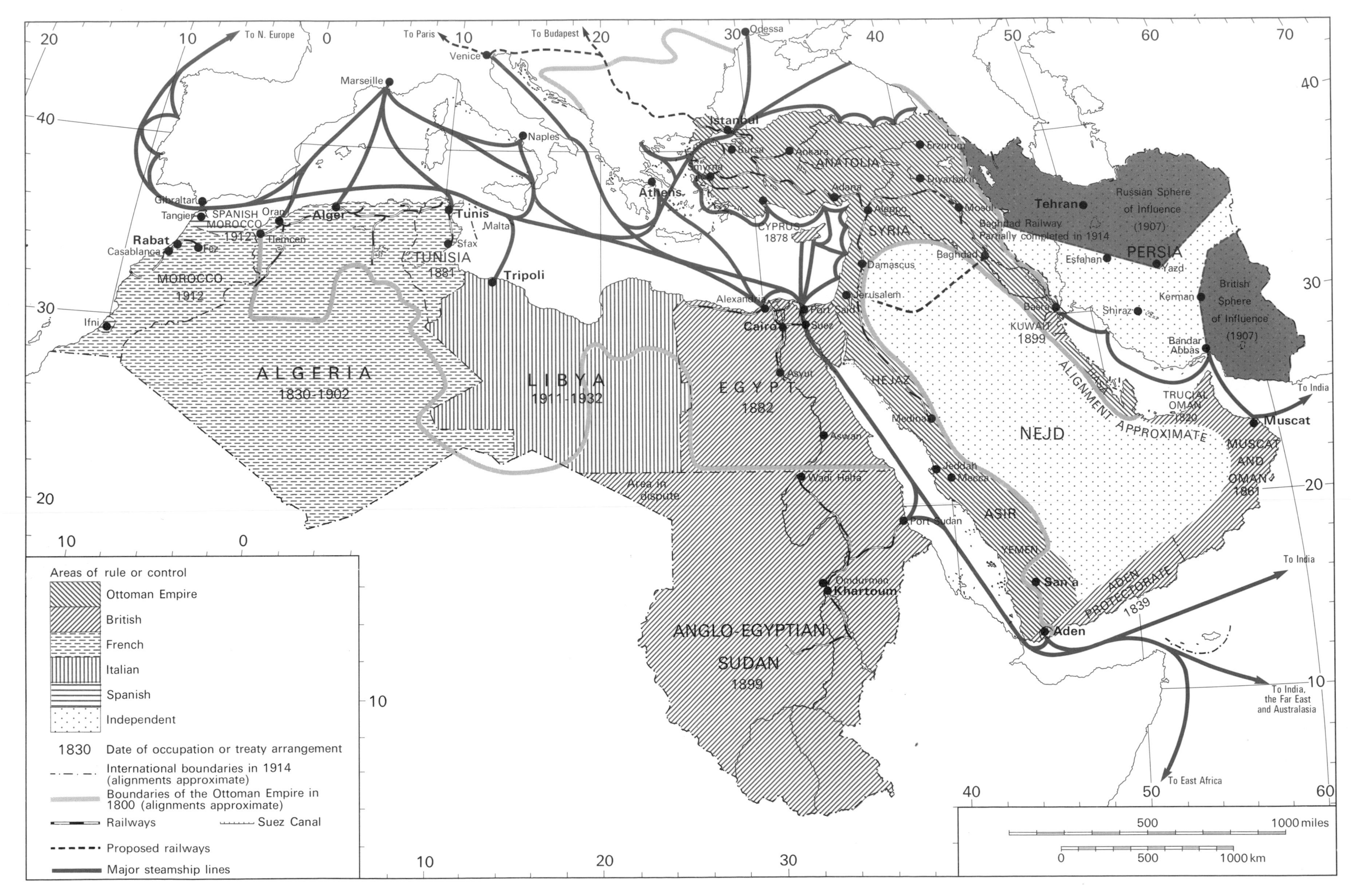

16 The region on the eve of World War I

The region on the eve of World War I

William Hale

Politically, the Middle East and Maghreb of 1914 could be divided into four parts: the Ottoman Empire, Persia, the North African states, and the centre and fringes of the Arabian peninsula. Political structures in each of these sections were highly diverse and need a brief description.

Although the Ottoman Empire had steadily lost territory during previous centuries, it still extended from the shores of the Arabian peninsula in the south to Anatolia and eastern Thrace, the European hinterland of Istanbul, in the north. The empire was the largest state in the region, but it was culturally heterogeneous and politically weak. In the Balkan Wars of 1912–13 the Ottomans had lost almost all their remaining territory in southeastern Europe. Nevertheless, ethnic Turks probably accounted for only about 45% of the empire's estimated total population of 21 million in 1914. Some 17% of the total were Christians (mainly Greeks and Armenians), with about 7% Kurds and the remaining 30% or so Arabs. Within the empire the government's power was shifting and uncertain. In theory, the Ottoman state was a centralised monarchy. Although the Young Turk regime which was installed after the revolution of 1908 had attempted to translate this theory into practice, the fact was that the government's control over outlying regions was often vestigial, and depended on the unreliable co-operation of local notables, usually Arabs or Kurds.

At first glance, Persia was closer to being a nation-state – an ancient kingdom which had preserved its independence for centuries, protected by its relative remoteness. On the other hand, the Shah's government was in most respects even weaker than that of the Ottoman Sultans. Then, as now, Persia contained substantial Turkic and Kurdish minorities. Its territorial integrity had been gravely jeopardised by the Anglo-Russian Convention of 1907, which divided the country into separate British and Russian spheres of influence.

North Africa, on the other hand, had been effectively taken over by the colonial powers. Most of Morocco, together with Algeria and Tunisia, was ruled over by France, but the constitutional status of the three territories differed. The formerly independent Sultan of Morocco was defeated by the French in 1912, and forced to accept a French protectorate. The administration of the northern zone of Morocco was then ceded to Spain, which also held a small enclave at Ifni, on the Atlantic coast. This pattern repeated that of Tunisia, where the Bey of Tunis (previously a nominal tributary of the Ottoman Sultan) had submitted to French protection in 1881. In practice, the role of both rulers was largely symbolic, and their countries were, in effect, governed by France.

In Algeria, the French established a straightforward colonial administration, with large-scale settlement by colonists from Europe. Italy attempted to implement a similar colonialist policy in Libya, which she seized from the Ottoman Empire in 1911, though with markedly less success. Egypt, meanwhile, had a highly ambiguous status. Nominally, it remained a part of the Ottoman Empire until the outbreak of World War I, when this legal fiction was ended. The government of the Khedives (who were themselves of Balkan, not Egyptian, descent) had formal responsibility for the internal government of the country but in practice the British, who conquered Egypt in 1882, controlled the levers of financial and military power. The Sudan, which had been under Egyptian rule until just before the British invasion but then broke away after the Mahdi's rebellion, was reconquered by an Anglo-Egyptian army in the late 1890s. In 1899 it technically became an Anglo-Egyptian condominium, but was to all intents and purposes under purely British rule.

On the southern and eastern shores of the Arabian peninsula there was a chain of petty states which were directly or indirectly controlled by Britain (in practice, by the British Government of India). The port and coaling station of Aden was administered directly from Delhi. In the hinterland, the British had separate agreements with the 23 local Sultans, sheikhs and tribal groupings of what was known as the Aden Protectorate. Further north and east, there were protectorate agreements with the Sultan of Muscat and Oman, the seven sheikdoms of what was then known as Trucial Oman, and the rulers of Bahrain, Qatar and Kuwait. The interior of Arabia was meanwhile a virtual no man's land, in which rival tribal leaders struggled for para-mountcy. Of these, the most important in 1914 were Abd al-Aziz Ibn Saud, the leader of the Wahabi sect, who recaptured the Saudi capital of Riyadh in 1902, and his rival, Sheikh Rashid of Hail, who ruled over the Shammar, to the north.

The map also shows the beginnings of railway construction in the region. Many of the principal cities – such as Alger, Tunis, Cairo, Damascus and Istanbul – were on or close to the Mediterranean coast, and had long enjoyed external communications by sea. It was only towards the end of the nineteenth century that modern transport links were constructed into inland areas. This development had crucial political as well as economic importance, since it enhanced governments' ability to control their own territories. The rail network was nevertheless sparse and patchy. In Morocco and Algeria the French constructed separate railways in the coastal belt, with an extension of the Algerian system into Tunisia. Libya was entirely without railways, but Egypt had a relatively dense network in the delta, with a branch running south along the Nile to Asyut. There was then a long gap in the network until the Sudan, which had one of the longest systems in the region. In the Ottoman Empire, the first direct Orient Express had rolled into Istanbul from Western Europe in 1888, and there were lines stretching inland from Izmir, on the Aegean coast. The empire's two major projects, the Baghdad and Hejaz railways, were barely completed by 1914. The Baghdad line still ended several hundred kilometres short of its destination, near the present Turkish–Iraqi frontier, and was broken by untunnelled ranges just north and east of Adana. These gaps were to prove a fatal handicap for the Turks in World War I, as they attempted to supply their armies in Iraq and Syria. The Hejaz railway was of great religious importance, since it connected the Holy Cities of Arabia to the rest of the empire, but it also represented an eleventh-hour attempt by the Ottomans to maintain their political hold over western Arabia. To the east, however, Persia was still untouched by railway development, underlining its relative isolation and backwardness at this time.

KEY REFERENCES

Issawi, C., *An Economic History of the Middle East and North Africa* (Methuen, London, 1982).

Kirk, G. E., *A Short History of the Middle East* (Methuen, London, 1948).

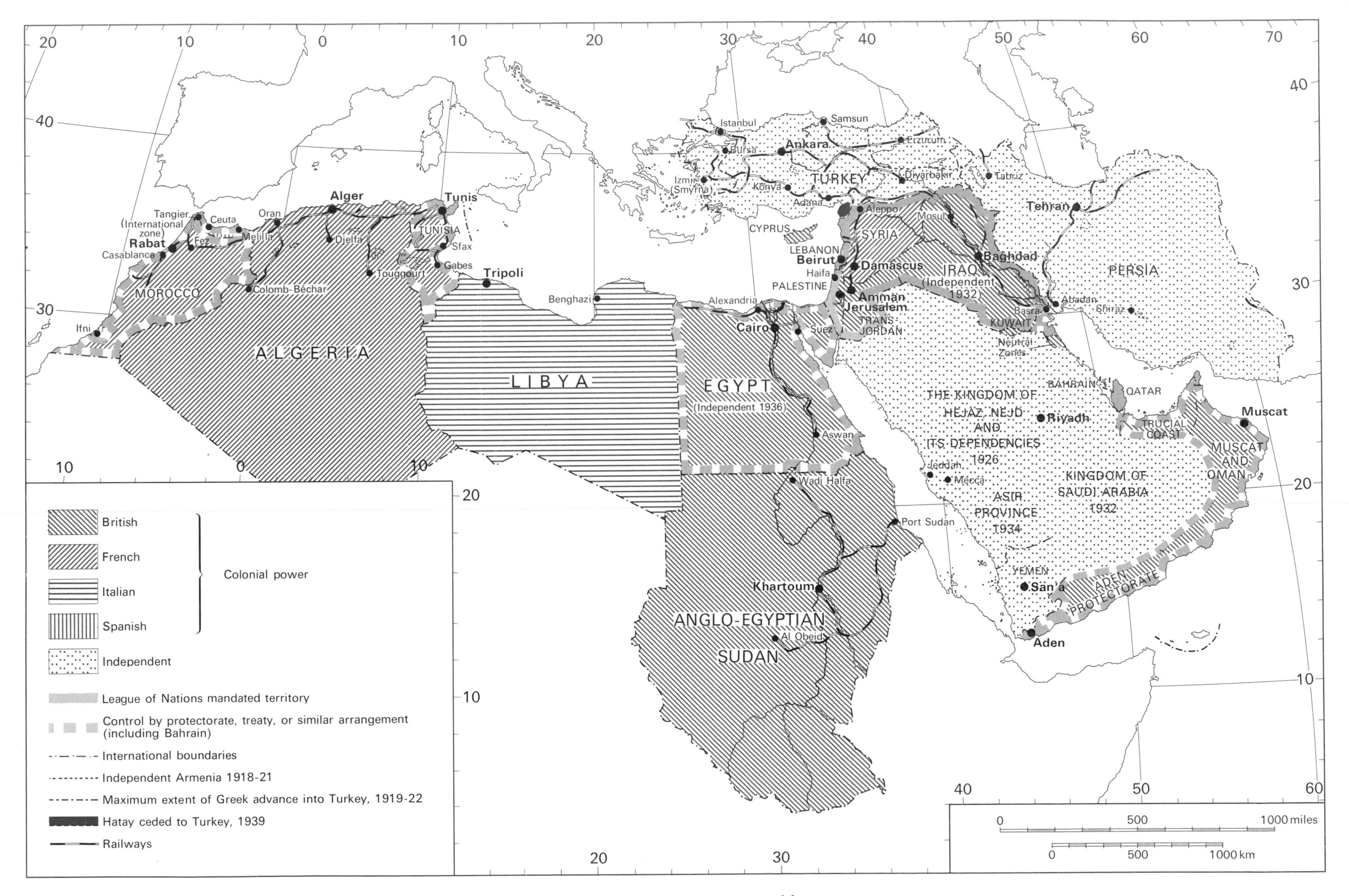

17 The region between the two world wars

Between the two world wars

William Hale

The map shows the fundamental political changes which had taken place in the region after the collapse of the Ottoman Empire in World War I.

During the war, the British and the other entente powers entered into a series of engagements which were to have a profound impact on the post-war political map. In 1916 British and French representatives negotiated the so-called Sykes–Picot agreement. This envisaged the establishment of direct French rule in Lebanon, northwestern Syria and southern Turkey, and British rule in southern Iraq, with an international regime in Palestine. Meanwhile, the British, who were attempting to foster an anti-Turkish rebellion in the Arab lands, entered into vaguely worded commitments to the Hashimi family, who were the hereditary governors of Mecca and nominal vassals of the Sultan. The Hashimis began their revolt in 1916, after receiving British promises of support for an independent kingdom in the Arab territories, with the disputed exclusion of Lebanon and western Syria. To add to the confusion, the declaration issued in 1917 by the British Foreign Secretary Arthur Balfour to the Zionist movement pledged support for the establishment of a 'national home for the Jewish people' in Palestine. Finally, Britain and France were forced to reconcile these conflicting commitments with President Woodrow Wilson's anxiety to secure the independence of small nations.

The pattern established after the war attempted to square the circle. Wilson's contribution was the establishment of the League of Nations, which in 1920 awarded supposedly temporary mandates for the government of Syria and Lebanon to France. Having separated Lebanon from the rest of Syria, the French further muddied the political waters by including in the Lebanese boundaries a number of Muslim areas, as a counterbalance to the Maronite Christian Lebanese. Britain was awarded mandates for Iraq and Palestine. As a consolation prize for the loss of the pan-Arab kingdom, the Hashimi Amir Feysal was made King of Iraq. The region east of the River Jordan was meanwhile separated from the rest of Palestine as the Amirate of Transjordan, under Feysal's brother, Abdullah. In Palestine, the British established a direct colonial authority. A trickle of Jewish immigration began which gathered pace during the 1930s. Legally, Iraq became an independent state in 1932, though it thereafter remained under substantial British influence. The mandate regimes were meanwhile continued in Syria, Lebanon, Palestine and Transjordan.

Equally dramatic changes were taking place in the Anatolian peninsula, the Turkish heartland of the old Ottoman Empire. Wartime agreements among the allies had envisaged the establishment of a Greek state in western Anatolia, and French- and Italian-ruled zones in the south (see Fig. 3). Greek troops landed in Izmir in 1919, but this triggered off a Turkish nationalist movement, led by Mustafa Kemal (Ataturk) who established his base in Ankara. By the late summer of 1921 the Greeks had advanced as far as the Sakarya river, some 80 km west of Ankara, but they suffered a devastating defeat in the following year. The French and Italians meanwhile abandoned their claims to Anatolian territory. The verdict of the battlefield was formalised by the Treaty of Lausanne of 1923, which established the Turkish state within its present frontiers (the sole subsequent territorial change was the transfer of Alexandretta, or Hatay, province from Syria to Turkey in 1939). Meanwhile, Ataturk had used his virtually undisputed national authority to abolish the sultanate and to lay the foundations of a modern, secularist republic.

To the south, the inter-war period saw the establishment of what is now the Kingdom of Saudi Arabia. With the collapse of Turkish rule, Ibn Saud was able to defeat his old rival, Sheikh Rashid of Hail, in 1921. He then turned his attention to the Hashimi kingdom in the Hejaz, which he conquered in 1925–6. Ibn Saud's 'Kingdom of Hejaz, Nejd and its Dependencies', set up in 1926, was re-named the 'Kingdom of Saudi Arabia' in 1932. The Saudi ruler fought a brief and successful war for the possession of Asir province with the now independent Imamate of the Yemen in 1934, thus extending his kingdom to its present size.

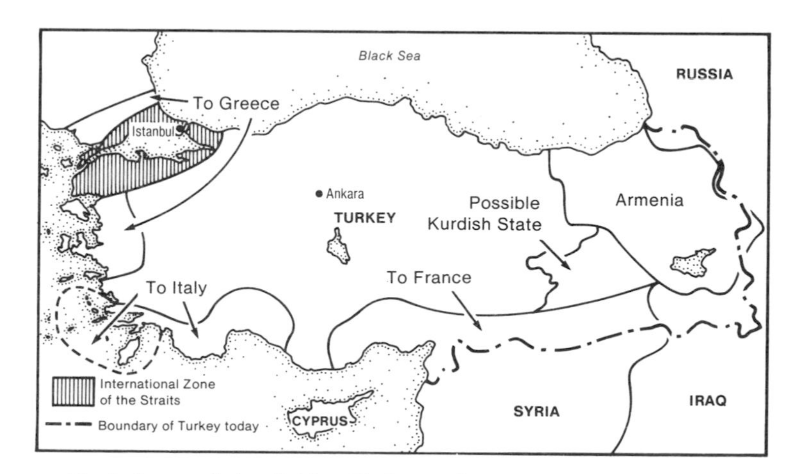

Fig. 3 Proposals for dividing Turkey under the Treaty of Sèvres, 1920.

In other parts of the region, there were some important political changes which, by their nature, cannot be shown on a map. During World War I, Persia had almost ceased to exist as an independent state, as large sections of its territory were periodically occupied by the British, Russians and Turks. In 1921–3, however, an effective national regime arose under the military leader Reza Khan. Reza re-established the authority of the government over the whole country and made himself Shah in 1925. In Egypt, meanwhile, the British faced severe challenges to their rule as a powerful Egyptian nationalist movement arose in the post-war years. The British gave ground in 1922–3 by making Egypt an independent constitutional monarchy. Their influence in the country nevertheless remained extensive, and Egypt's exact status remained in dispute until an Anglo-Egyptian Treaty was signed in 1936.

In the other North African countries, nationalist movements were also stirring. In Morocco, Abdulkerim's rebellion in the Rif was only brought to heel by a massive Franco-Spanish campaign in 1926. Tribal resistance in other regions was stubborn, if unco-ordinated, and lasted until 1934. Meanwhile, nationalist independence movements were gathering strength in Tunisia and Algeria, in the shape of the Neo-Destour and Algerian Popular Parties respectively.

As the map shows, the inter-war years saw a significant extension of railways in the region, with important implications for national and international integration. The Istanbul–Baghdad Railway was completed, with a narrow-gauge extension to Basra. In Iran, the Trans-Iranian Railway, linking Tehran directly with both the Caspian and the Gulf, was the major physical achievement of Reza Shah's reign. In Turkey, the network was extended eastwards to Diyarbakır, Erzurum and the Russian frontier. The Hejaz railway, however, was severely damaged by the wartime fighting in Arabia, and now stopped short at Amman. In Egypt, rails ran south as far as Aswan, while in French-ruled North Africa there was now a continuous rail link from Tunis in the east to Casablanca in the west.

KEY REFERENCES

Issawi, C., *An Economic History of the Middle East and North Africa* (Methuen, London, 1982).

Kirk, G. E., *A Short History of the Middle East* (Methuen, London, 1948).

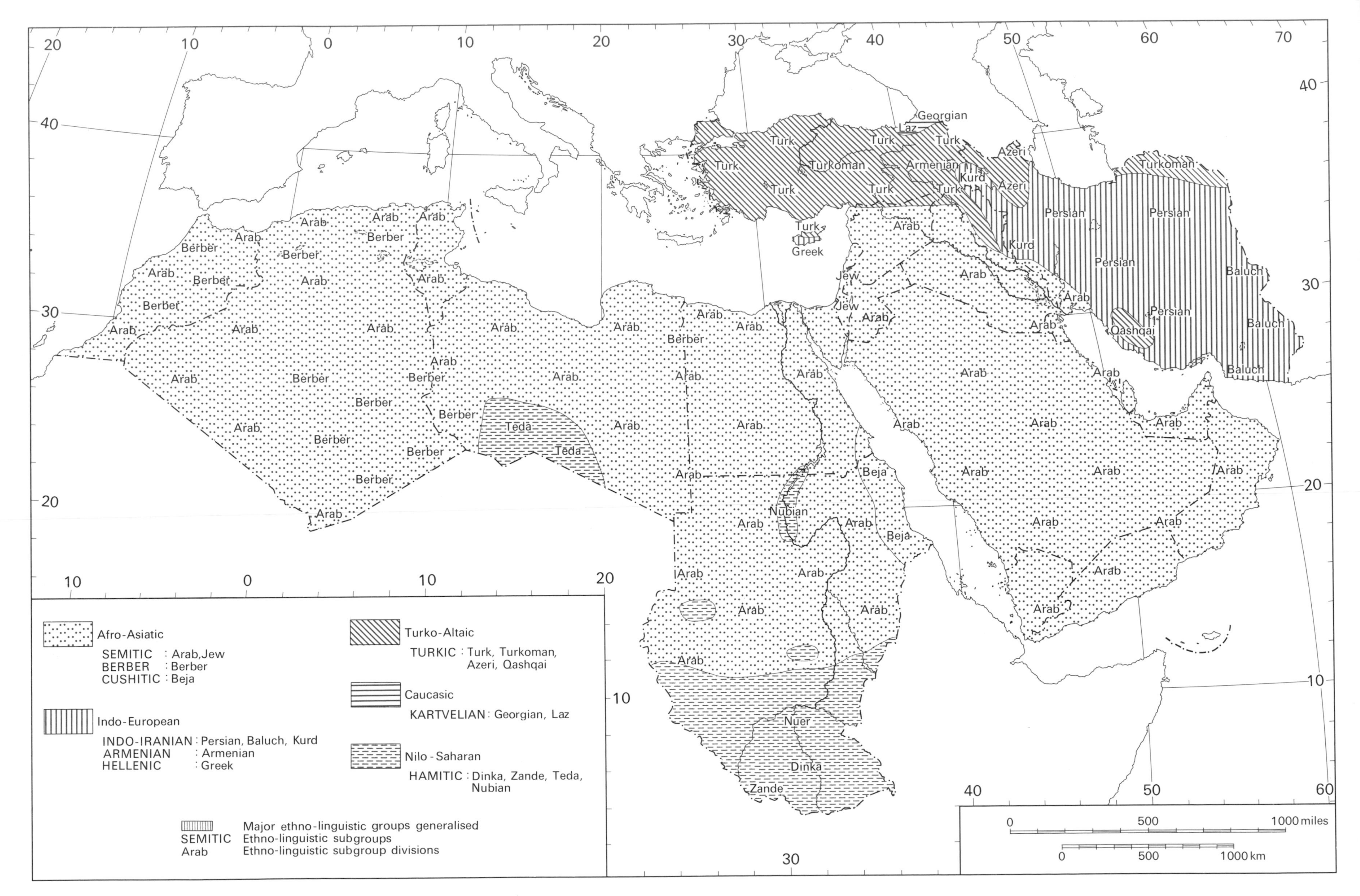

18 Ethno-linguistic regions

Ethno-linguistic regions

Michael Drury

An ethno-linguistic map is at once a fascinating and a potentially dangerous document. On the basis of such maps the boundaries of nation-states have been established and justified, at least in Europe, for 200 years. And so powerful has been the force of the argument linking ethno-linguistic distinctiveness with a national consciousness and a right to political independence that contemporary distributions continue to form the basis for demands for boundary modification, for kindling ambitions for new states, and for political manipulation in order to conceal 'true' distributions. All of these facets are found within the Middle East and North Africa, but the map both reveals and disguises the ethno-linguistic realities.

What is concealed should be dealt with first; the dangers of mapping spatial distributions on the regional scale are both obvious and subtle. Certainly, since most of the very sparsely populated areas are classified as 'Arab', the inference that this group is as dominant numerically and politically as its spatial supremacy suggests is too easily drawn. Similarly, the scale cannot cope with two distinct levels of complexities which, in reality, are of critical importance: one is the detailed variations occurring within the more densely populated 'tribal' areas, above all in southern Sudan, where more linguistic groups are found than in the whole of the rest of the Middle East; the other is the marked heterogeneity of many areas, including some of the major cities. The Greeks of Istanbul, for instance, are as numerous as the grossly depleted Armenian population of Turkey, yet, occupying but a spatial pin-point, they can achieve no place on the map. Finally, these distributions represent a compromise between 'historic' core areas of certain groups and the *de facto* supremacy of others: thus, Israel is shown as exclusively Jewish, although 18% of its population is Arab. Similarly such names as 'Laz' and 'Armenian' in eastern Turkey represent only the residuals of once much larger populations.

What does the map reveal? Above all, it shows the ubiquitousness of the Arabs throughout most of the Middle East and North Africa with the important exceptions of Turkey, Iran and southern Sudan. 'Arab' is used in this context simply, and properly, to describe all those whose mother-tongue is Arabic. As such, the distribution represents the consequences of a dynamic and very rapid expansion of the Arab population, directly associated with the first century after the rise of Islam, during which time they spread from southern Arabia to embrace the whole of North Africa and most of the lands between Arabia and the Indus (Map 14). The resistance of the less well-organised peoples they encountered remains clear in contemporary distribution patterns, above all in North Africa. Here, Berber groups are found in all five Mediterranean states, being particularly strong in Morocco where, almost exclusively in mountainous environments, their own loose confederative system of tribal self-rule remains, 1300 years later, in uneasy alliance with the central governmental structures.

Within North Africa, however, only in Sudan is the dominant Arab political presence not reflected in equal degrees of either spatial or numerical dominance. Here almost 600 distinctive ethnic groups have been identified and only 40% of the population regards itself as 'Arab'. The rest include the Nubians, many of whom were displaced as Lake Nasser rose following the building of the Aswan Dam (Map 12); and the Beja, pastoral nomads, some of whom have been successful in diversifying their income by utilising their location astride Sudan's road and rail outlets to the sea by engaging in 'land piracy'. The many Negroid and Nilotic groups in the south and southwest, among them the Dinka, Nuer and Zande, are particularly important. These peoples draw Sudan towards the rest of Africa rather than the Middle East, and have resisted social and political integration.

In contrast, the near ethno-linguistic homogeneity of Egypt reminds us that 'Arab' does not equal 'Muslim', for here at least 10% of the Arab population is Coptic Christian.

The Arabian peninsula proves to be aptly named: as their area of origin, it remains the core of the Arab world but, despite huge economic and political growth, it contains less than 10% of the Arab population. Of these, half live in the impoverished Yemens where tribal affiliation has tended to be more dominant than the feeling of being Yemeni and the latter has frequently been more important than the sense of being 'Arab'. It is, however, only when one reaches the northern fringes of 'Arabia' that the deeper ethno-linguistic complexities of the Asiatic Middle East emerge.

Above all, here is Israel. The fact that it is not exclusively Jewish has already been noted; in addition, over a million Palestine Arabs live in exile immediately beyond its present borders. The Israeli government has used the reborn Hebrew language as an integrating force amongst its Jewish population (82%), an example of the importance of the linguistic factor in moulding a 'nation'. Conversely, neighbouring 'Arab' Lebanon has shown in its recent history the dangers of reliance on ethno-linguistic unity when beneath lie deeply rooted religious and economic differences (Map 57). The limits of the Arab realm are found close to the northern borders of Syria and Iraq. The non-coincidence between peoples and boundaries in these areas continues to give scope for conflict, whether it be Syria's dormant wish to secure the Iskenderun 'finger' of Turkey, or Iraq's more active wish to 'free' the Arabs of southwest Iran. Here also, straddling these four countries, are the Kurds. They, like the Berbers, maintain an essentially tribal structure which may have led to their survival, but has not encouraged their political evolution. Far more numerous than their spatial distribution suggests, they are the fourth largest ethno-linguistic group of the region, numbering some 10 to 12 million. The Kurds have had their aspirations for autonomy thwarted by their neighbours who have used them both as pawns in local international relationships and as targets for the process of 'ethnic purification' in which all have engaged to some degree.

Turkey and Iran account for some 85% of the region's 100 million non-Arabs. Both, but especially Turkey, have pursued policies to achieve ethno-linguistic homogeneity. Despite that, in Iran only 55% regard Farsi (Persian) as their native tongue. A further 25% speak related Iranian languages, Baluchi being one of the spatially more important. Far more significant are the 5 million speakers of Azeri, a Turkish tongue, located in the potentially unstable northwest Iran. No such linguistic minority strength remains in Turkey. A century ago 1.6 million Armenian-speakers lived within Turkish Armenia; today, the number is negligible. With the exception of the Kurds, who have proved remarkably resistant to 'Turkification', the other minor linguistic groups of Turkey have dwindled to relative insignificance. Cyprus is also highly significant in an ethno-linguistic sense. Here, since 1974 the Turkish-speaking population has been separated from the Greek-speaking population by an armistice line, where previously they intermingled. This line is proving as rigid a frontier as any within the Middle East. It is to be hoped that this most westernised corner of the Middle East does not set too many precedents for the resolution of dynamic ethno-linguistic patterns.

KEY REFERENCES

Clarke, J. I. and Fisher, W. B. (eds.), *Populations of the Middle East and North Africa* (University of London Press, London, 1972).

McEvedy, C. and Jones, R., *Atlas of World Population History* (Penguin Books, Harmondsworth, 1978).

Minority Rights Group, *The Armenians*, no. 32 (M.R.G., London, 1981).
The Baluchi, no. 48 (M.R.G., London, 1984).
Cyprus, no. 30 (M.R.G., London, 1984).
Israel, no. 12 (M.R.G., London, 1981).
The Kurds, no. 23 (M.R.G., London, 1981).
Lebanon, no. 61 (M.R.G., London, 1981).
The Palestinians, no. 24 (M.R.G., London, 1982).

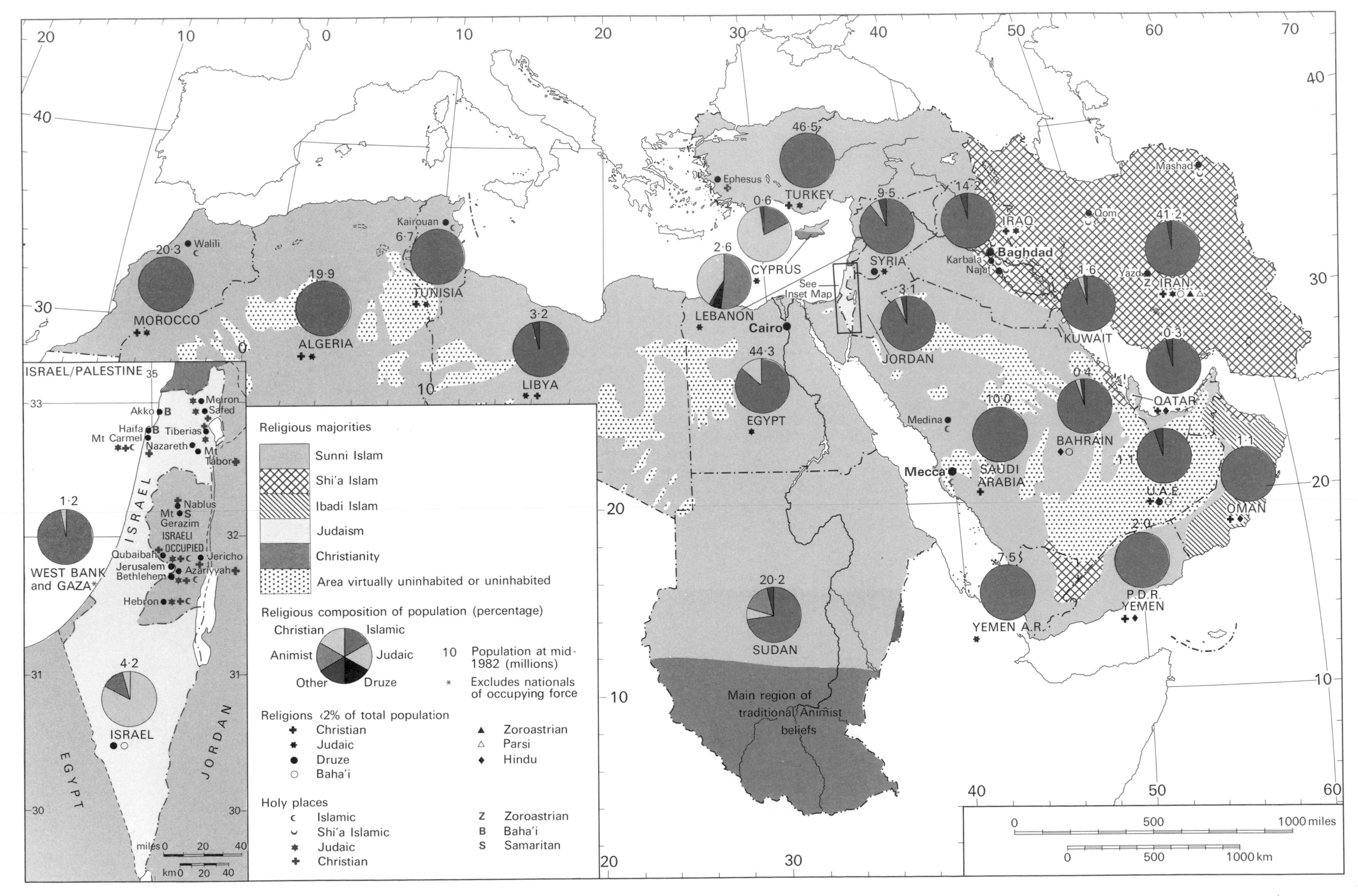

Religious majorities
Sunni Islam
Shi'a Islam
Ibadi Islam
Judaism
Christianity
Area virtually uninhabited or uninhabited
Religious composition of population (percentage)
Christian
Islamic
Animist
Judaic
Other
Druze
10 Population at mid-1982 (millions)
* Excludes nationals of occupying force
Religions ‹2% of total population
Christian
Judaic
Druze
Baha'i
Zoroastrian
Parsi
Hindu
Holy places
Islamic
Shi'a Islamic
Judaic
Christian
Z Zoroastrian
B Baha'i
S Samaritan
MOROCCO 20·3
ALGERIA 19·9
TUNISIA 6·7
LIBYA 3·2
EGYPT 44·3
SUDAN 20·2
TURKEY 46·5
CYPRUS 0·6
LEBANON 2·6
SYRIA 9·5
IRAQ 14·2
JORDAN 3·1
KUWAIT 1·6
IRAN 41·2
QATAR 0·3
BAHRAIN 0·4
SAUDI ARABIA 10·0
U.A.E. 1·1
OMAN 1·1
YEMEN A.R. 7·5
P.D.R. YEMEN 2·0
Walili
Kairouan
Ephesus
Cairo
See Inset Map
Baghdad
Karbala
Najaf
Qom
Mashad
Yazd
Medina
Mecca
Main region of traditional Animist beliefs
ISRAEL/PALESTINE
Akko
Haifa
Mt Carmel
Meiron
Safed
Tiberias
Nazareth
Mt Tabor
Nablus
Mt Gerazim
ISRAELI OCCUPIED
Qubaibah
Jerusalem
Bethlehem
Jericho
Azariyyah
Hebron
WEST BANK and GAZA* 1·2
ISRAEL 4·2
EGYPT
JORDAN
miles 0 20 40
km 0 20 40
0 500 1000 miles
0 500 1000 km

19 Religion

Religion

John Chitham

The Middle East is the birthplace of the world's three great monotheistic religions: Judaism, Christianity and Islam. Over the centuries these have splintered into different sects and have been influenced by different philosophies such as Greek thought, Zoroastrianism and Hinduism. This has resulted in a complex mosaic of beliefs as each wave of religious revelation dominated but did not eradicate the last. Thus the region is the heartland of Islam, yet several minority religions and sects remain, often situated in curious locations after long and turbulent histories. Through the centuries and to the present day the religious life of the region has brought enrichment and unity, and conflict and persecution.

Statistics on religious confession in the Middle East and North Africa are controversial and those quoted below can only be treated as estimates. Nevertheless the map clearly shows that, apart from Israel and Cyprus, every country in the Middle East and North Africa has a Muslim majority. The former Christian majority in the Lebanon was reduced to a minority during the 1970s or before.

The Sunni branch of Islam is the most numerous, dominating North Africa and the Arabian peninsula, totalling almost 200 million adherents. Shi'ism is in a majority in Iran and Iraq, where there is a combined total of some 50 million Shi'ites, and has important minority communities in Lebanon, Yemen A.R., Syria and Oman. The location of Shi'ites today still follows the pattern established by the political divisions in the first century of Islam. Ibadism, another sect which began in the same political turmoil, is now only important in Oman, where it is the national religion. Semi-secret sects have also come out of Islam, including the Druzes and Yazidis. Of these the Druzes are the most important, having been founded in the eleventh century A.D. and surviving tenaciously in mountain strongholds in Israel, Lebanon and Syria.

Before the formation of the state of Israel the Jews were a large and old-established minority in most countries of the Middle East and North Africa. Since 1948 there has been large-scale migration from these countries to Israel, so that only a small number remain, chiefly in the states of North Africa, the Levant, Iraq, Iran and Turkey. The four million or more Jews in Israel are divided between the Sephardi and Ashkenazi, the former being mainly 'Oriental' Jews from the Middle East and North Africa, and the Ashkenazi mainly from Europe.

Christianity has innumerable sects in the Middle East. The origin of these divisions was both political and theological, the principal divisions occurring in the fifth and eleventh centuries A.D. It is in the Levant that the ancient church has survived in the largest numbers. The Copts in Egypt are most numerous with between four and six million adherents. The Greek Orthodox are the most important in Syria, Jordan and the Occupied Territories, while the Maronites and Greek Catholics (both owing allegiance to the Pope in Rome) are most numerous in Lebanon and Israel respectively. In total in the Levant there are over 500,000 Greek Orthodox and about one million Maronites, the latter almost exclusively in Lebanon – Iraq and Iran also have substantial numbers of Christians, especially Chaldean Catholics and Armenian Orthodox. Cyprus is divided between the Greek Orthodox majority in the south and the Turkish Muslims in the north. The north of Sudan remains strongly Muslim, but in the south most people are either Christian or Animist. A new element in the distribution of religious groups has been created by opportunities for employment as a result of increased oil revenues, especially since 1973, taking for example Egyptian Copts to Libya and Hindu workers to the Gulf states (Map 24).

Religious centres abound in the Middle East. Many of these are centres of learning, but most of the important schools are also at the site of places of pilgrimage or of historical significance to the religion concerned. In Islam, Mecca is the most significant centre for pilgrimage. The Quran commands that the annual Haj to Mecca is the duty of every Muslim once in a lifetime (Table 4). Medina, the city from which Muhammed set out to conquer Mecca, is the second city of Islam, with Jerusalem the third. The Dome of the Rock marks the traditional site where Muhammed left to visit heaven. Other Sunni centres include Al-Azhar University in Cairo, Kairouan in Tunisia, and Walili in Morocco. Shi'ite centres are numerous, due to the Shi'ite's reverence for martyrs and their shrines. The most significant are at Najaf, burial place and shrine of Ali, revered by Shi'ites and son-in-law of Muhammed; Karbala, site of the Sunni/Shi'ite battle where Husayn, son of Ali, died and was enshrined; Mashad, site of the shrine of the eighth Imam in Iran; and Qom, shrine of the eighth Imam's sister and seat of the Ayatollah Khomeini. Also in Iran is Yazd, where the largest remnant of the Zoroastrians reside.

The densest clustering of religious centres probably occurs in the Israeli-occupied West Bank. This is the geographical focus of Judaism and Christianity, while Muslims also revere the sites of the deeds of many commonly accepted prophets, as well as Jerusalem. The four holy cities of Judaism are Jerusalem, Hebron, Tiberias and Safed. Safed is also the centre of the mystical Jewish sect of Kabbalism, while Meiron contains the tomb of the Kabbalist Rabbi Simeon ben Yohai. Christians regard the sites associated with the actions of Jesus as specially significant. These include Bethlehem (his birth); Jericho (baptism); Nazareth (early life); the region of Lake Tiberias (miracles); Mount Tabor (transfiguration); Jerusalem (death and resurrection); and Qubaibah (Emmaus road sermon). Two other religions have their centres in Palestine. The Samaritans built their temple at the foot of Mount Gerazim when they and the Jews returned from exile in the fourth century B.C., and the Bahai have the shrine of the Bab (their founder) at Akko, as well as their Universal House of Justice in Haifa, despite having their origins in Iraq. Jerusalem symbolises more than any other site the division and indivisibility of the faith of the region, as the capital of King David and site of the Temple, the place of the death of Jesus, and the third holy city of Islam.

Table 4. *Overseas pilgrims to Mecca*

Period	Average annual number
1936–39	52,117
1940–44	39,519
1945–48	65,768
1949–53	134,131
1954–58	217,247
1959–63	224,266
1964–68	317,318
1969–73	513,967
1974–78	1,025,486
1979–83	881,867

Sources: Various.

KEY REFERENCES

Arberry, A. J. (ed.), *Religion in the Middle East*, vols. 1 and 2 (Cambridge University Press, Cambridge, 1969).

Encyclopaedia of Islam, vols. 1–3 (A–Iran), new edn, vols. 4–5 (Iran–Mahi) (E. J. Brill, Leiden, 1979–86).

Serjeant, R. B., 'Religions of the Middle East and North Africa', *The Middle East and North Africa 1983–84* (Europa Publications, London, 1983), pp. 24–31.

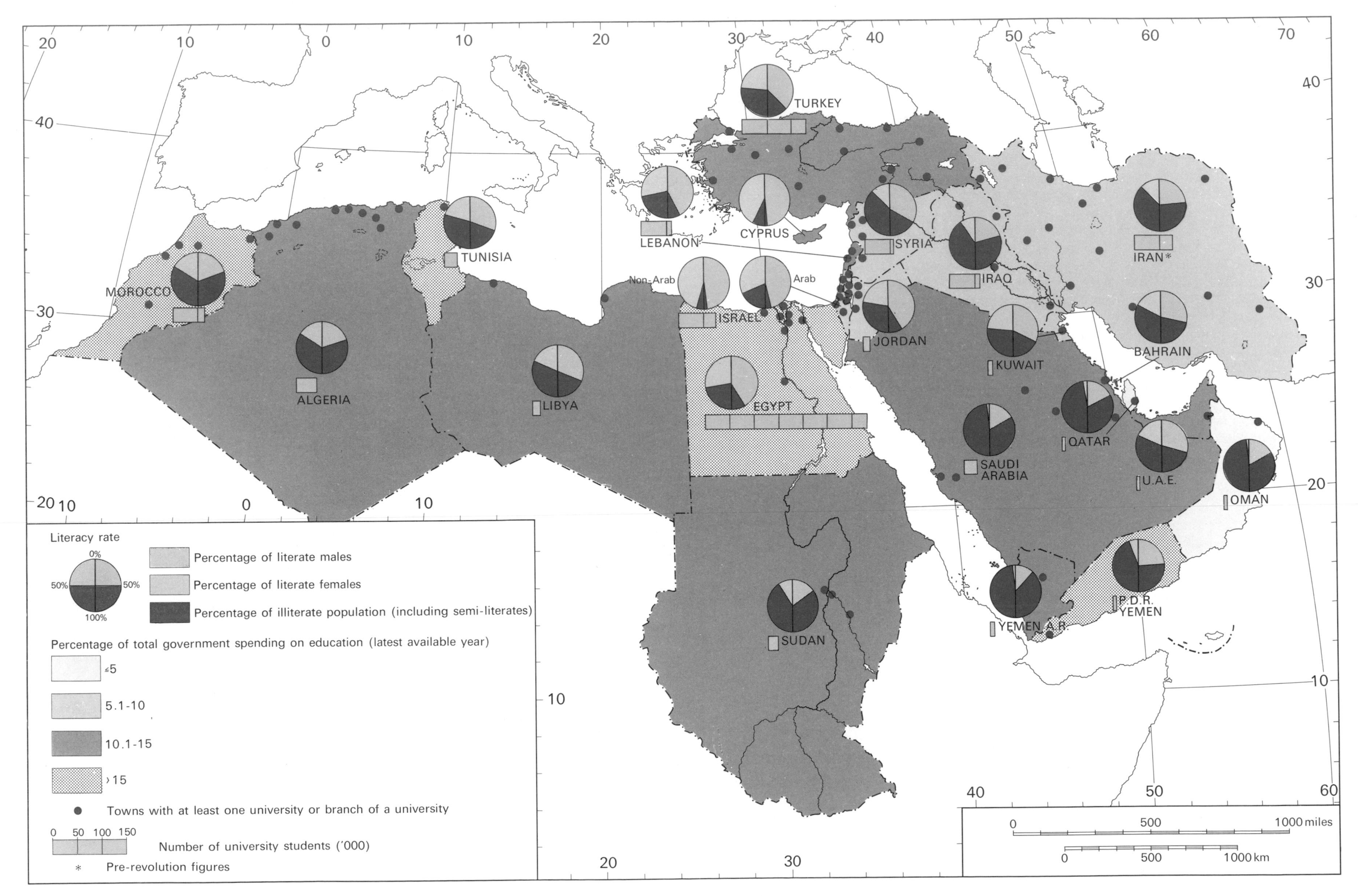

20 Literacy and learning

Literacy and learning

Bill Williamson

Diversity is what characterises the patterns of education and culture in the Middle East. It is manifest in different models and levels of social and economic development and the amount and quality of resources, both physical and human, available for educational purposes. Such differences, in their turn, reflect diversity in the cultural traditions of Middle Eastern societies and the nature and timing of their most significant contact with Western civilisation.

Islam is central to an understanding of education in the area. It should not be forgotten, however, that Ottoman Turkish and Persian cultural traditions as well as those of Christian and Jewish minorities have each played a part in shaping educational and cultural values in the area as a whole (Maps 14–19). In the period from the end of the eighteenth century onwards, the intellectual and cultural influences of Western civilisation penetrated the Middle East. The consequences of this have been profound and how the impact of the West can be reconciled with the cultural values of the Middle East is one of the most urgent and difficult problems educators in the region have to face.

It is enjoined on the Muslim faithful to seek knowledge. The Prophet is believed to have said, 'The ink of the learner and the blood of the martyrs are of equal value in the sight of heaven' and that 'A father can confer upon his child nothing more valuable than a good education.' This religious stimulus to scholarship explains much of the early vitality of Islamic civilisation and the importance which it attaches to literacy.

The geographical expansion of Islam from the seventh century onwards was accompanied by remarkable developments in science, mathematics, medicine and, of course, in the religious sciences. Arab and Persian scholars drew liberally on Greek, Persian and even Indian scholarship. Arab translations of classical Greek texts filtered through Italy, Spain and the Levant were part of that cultural flowering in Europe known as the Renaissance. By the time of the great medieval scholar, Ibn Khaldun, this tradition was already in decline. The consolidation of the Faith and its traditions by the doctors of Islamic law, the *ulema*, and their mistrust of intellectual activity beyond a strict religious boundary are two factors often adduced to explain this.

This civilisation nevertheless developed schools (*kuttabs*) and colleges (*madrasahs*) through which to convey the tenets of the Faith throughout that vast geographical area from the Atlantic coast of Africa to the Hindu Kush. The mosque of Al-Azhar in Cairo, founded in A.D. 969, is said to be one of the oldest universities in the world and has a central place in the scholarship of Islam as a whole.

The tension between religious authority and speculative enquiry is paralleled by another, that between the town and the countryside. This is of great significance for education within the Muslim community (the *umma*). The literate traditions of the urban *ulema* have always been different from those of the people of the countryside whose 'folk Islam' was entangled with pre-Islamic beliefs and superstitions and the mystical practices of Sufism. This historical fact still bears heavily today on the statistics of literacy and the distribution of educational opportunities of a formal academic kind. In general, the rural areas of all Middle Eastern societies have much poorer educational provision and higher rates of illiteracy than urban areas. It is a difference, too, which explains the contrast in the education of men and women; throughout the Middle East, with the exception of Israel, Cyprus and the Lebanon and, to a lesser extent, Egypt, female illiteracy rates are significantly higher than those for men.

Western influence in the Middle East increased dramatically from Napoleon's invasion of Egypt in July 1798. This prompted defensive modernisation throughout the Ottoman lands, in which the development of specialist schools, often staffed with foreign experts, played a key role. General elementary education was neglected throughout the nineteenth century. The Maghreb was brought under French colonial influence and cultural control. Egypt and the Sudan and much of Persia were under British control. Foreign influence in the area was to encourage the growth of education and educational aspirations of a Western type and to hold back the general education of the mass of the people.

The themes around which the educational experience of the region in the twentieth century and particularly since World War II should be discussed are those of modernisation and national independence and persisting patterns of underdevelopment. Throughout the region education has expanded at all levels. This expansion has been a goal and a precondition of modernisation. Countries like Egypt, Turkey and Syria have had a much longer experience of this than others such as Saudi Arabia or the Gulf states. The late developers have therefore had to rely considerably on the importation of foreign expertise and personnel. This has given the developed societies yet more opportunities for influence in the region and has encouraged massive labour migration among educated people. Egypt supplies the whole Arab world with many doctors, engineers and teachers. Small countries like Qatar accommodate expatriate populations more numerous than their own nationals.

The pace of modernisation in countries like Iran and Saudi Arabia and the oil-rich states of the Gulf has been accompanied by great social and cultural changes. Traditional values and life styles persist uneasily alongside those of the modern world and can come into fundamental conflict with it. The Iranian revolution is a case in point. So too is the growth of Islamic revitalisation movements throughout the region. It is not really surprising that these conflicts find their most acute expression in the field of education. There is a great concern in Saudi Arabia about the pervasive values of what they regard as the modern, 'godless civilisation' of the developed world, and students in Iran have been at the forefront of the revolution.

The upward pressure for modern education is, however, irresistible. Even though it does contribute to new kinds of social inequalities in the societies of the region it is an indispensable precondition of further development in the economic and social life of the region as a whole.

KEY REFERENCES

Szyliowicz, J. S., *Education and Modernization in the Middle East* (Cornell University Press, Ithaca, 1973).

Tibawi, A. L., *Islamic Education: Its Traditions and Modernization into the Arab National Systems* (Luzac and Co., London, 1972).

Williamson, Bill, *Education and Social Change in Egypt and Turkey* (Macmillan, London, 1986).

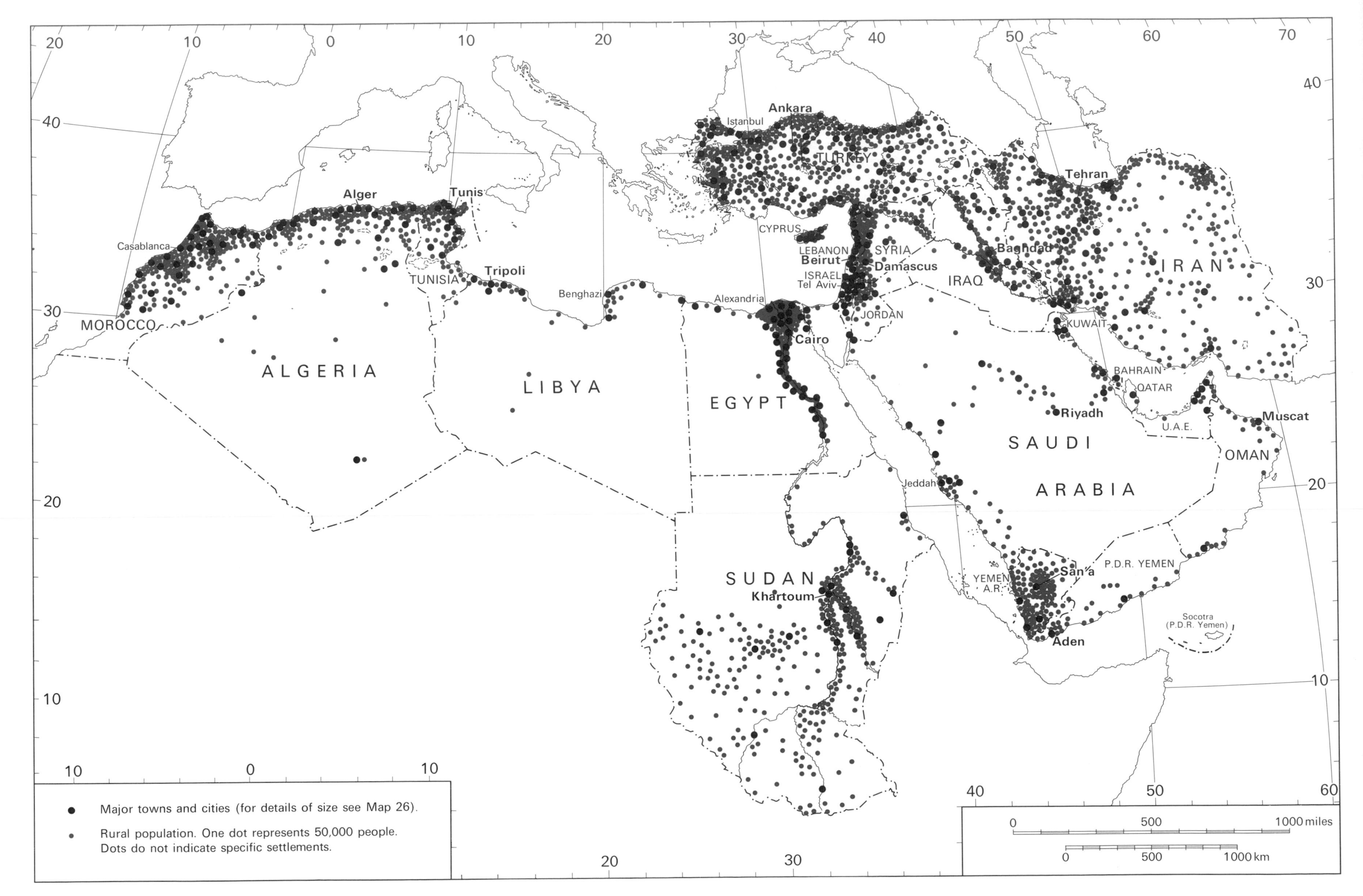

21 Population distribution

Population

John Dewdney

This topic is covered by a set of three maps (21–23) which portray population distribution, population density and population change respectively. Distribution and density are, of course, closely interrelated, not least because they are derived from the same set of data, and to discuss these sequentially would confuse rather than enlighten the reader. Population change could be treated as a separate topic but, since change in population numbers implies changes in distribution and density, this aspect is also closely linked to the other two. Consequently a single text is presented to cover the information provided by all three maps and Table 5 also covers all three aspects.

Population distribution and density

Covering an area of some 14.8 million km^2 – roughly one-tenth of the earth's land surface – the Middle East and North Africa had a total population of about 270 million in 1983, little more than 5% of the world total. Thus the overall population density for our region, about 18 per km^2, is little more than half the current world average of 33 per km^2. However, this not inconsiderable population – roughly equivalent to that of the U.S.S.R. and nearly five times that of the U.K. – is very unevenly distributed both between and within the 22 states into which the Middle East and North Africa are divided, and the region contains some of the world's most densely populated districts as well as some of its most extensive empty areas, the two extremes often occurring within the boundaries of a single state.

As Fig. 4 and Table 5 indicate, national populations cover a wide range from less than half a million in some of the micro-states of the Arabian/Persian Gulf to more than 40 million in Turkey, Egypt and Iran. The latter three countries, while accounting for less than a quarter of the region's land area, contain just over 50% of its population. The next three countries in order of population size – Morocco, Sudan and Algeria – each have around 20 million inhabitants and together account for just below one-quarter of the total, and a further three with populations of 9–15 million – Iraq, Syria, Saudi Arabia – another 13%. This leaves about 34 million people distributed among the remaining 13 states, which together have about 13% of the population living in 8% of the land area and populations ranging from about 7 million (Tunisia, Yemen A.R.) to as little as a quarter of a million (Qatar).

These variations in the size of national populations bear little relation to the territorial extent of the various countries, as Table 5 indicates. Libya, for example, with an area of 1,760,000 km^2, has a population smaller than that of Israel, which has an area of only 21,000 km^2; Saudi Arabia is nearly three times the size of Turkey but has only one-fifth of that country's population. These discrepancies are related to regional variations in population density, details of which are illustrated in Map 22. Such variations are related primarily to the physical environment as it affects the possibilities for agriculture, a major element being the matter of water supply.

While the average population density for the Middle East and North Africa is about 18 per km^2, few areas have densities close to this figure; most are either well above or well below the regional average. Considerable contrasts in density can be seen even at the level of the countries and major subregions identified in Table 5. The density of the North African section as a whole (14 per km^2) is about two-thirds that of the Southwest Asian section (23.6) and each of

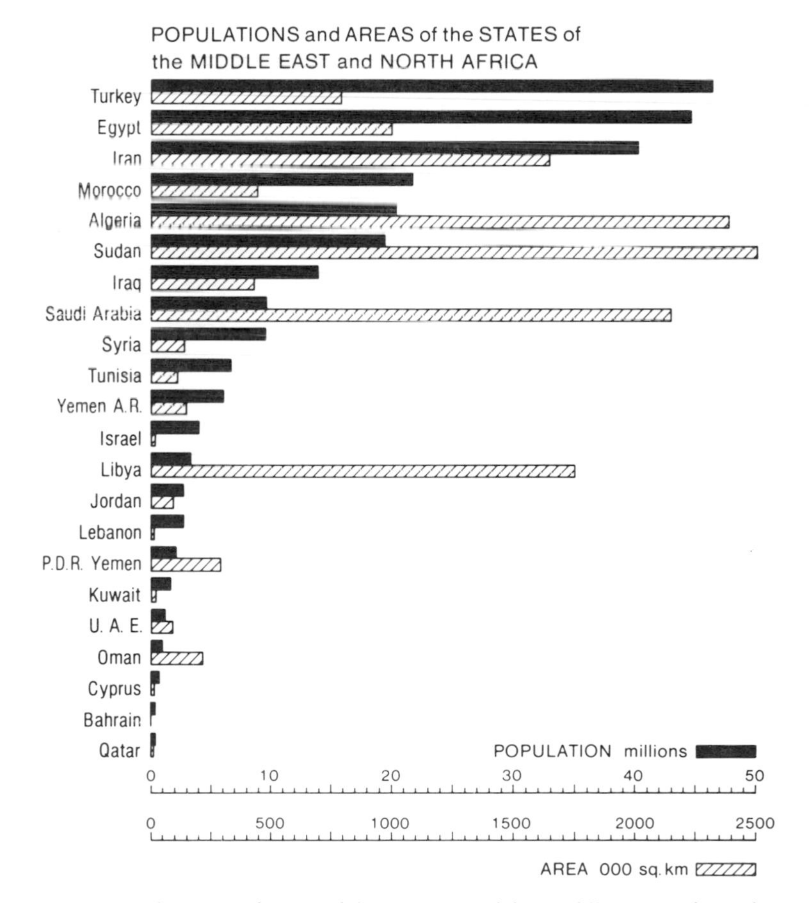

Fig. 4 Populations and areas of the countries of the Middle East and North Africa.

these major divisions has pronounced internal variation. In the Maghreb, Tunisia (41.4) and Morocco (45.3) have densities 5 or 6 times that of Algeria (8.6), and in Northeast Africa Egypt's average density (44.6) is nearly 25 times that of Libya. In the Southwest Asian section, the most obvious contrast is that between the Arabian peninsula (7.5) and the more northerly areas – the northern tier (38.6) and the Levant/Mesopotamia (46.0). In the latter two divisions all countries have densities well above the average for the Middle East and North Africa as a whole; the Levant-Mesopotamia contains two of the most densely populated countries in the whole region, Lebanon (260.0) and Israel (195.2). In the Arabian peninsula, the low overall density is due mainly to the very low densities in Saudi Arabia (4.3) although Oman (4.0), P.D.R. Yemen (6.5) and the United Arab Emirates (13.1) are also well below average. The remaining countries have relatively high densities, most notably Bahrain, which has over 600 persons per km^2.

More significant than these contrasts at the national level, however, are the even more pronounced density variations within individual states, the general pattern of which is indicated in Map 22.

Areas of highest density, with more than 100 persons per km^2, are of two main kinds. The highest densities of all are associated with areas of intensive irrigated agriculture along major river courses. Outstanding among these is the main settled zone of Egypt, along the Nile – which contains the great bulk of that country's inhabitants – and its extension upstream into northern and central Sudan. Densities exceed 500 per km^2 throughout most of this narrow riverine belt and reach as high as 2000 per km^2 in parts of the Nile delta. In central Iraq, densities of several hundred per km^2 are recorded in the valleys of the Tigris and Euphrates. A number of smaller pockets of very high density associated with irrigated agriculture occur elsewhere, for example around Damascus and Riyadh. A second group of very high density areas occurs where precipitation is sufficient to support fairly intensive rain-fed agriculture with the occasional assistance of irrigation. Such areas are mainly coastal: the eastern part of the Turkish Black Sea coast, the Caspian coast of Iran, the eastern Mediterranean or Levant coast and small sections of the Maghreb coast in Morocco, Algeria and Tunisia. Pockets of very high density too small to be indicated on the map also occur in oases scattered over otherwise very thinly settled or empty desert areas.

Areas with densities of 40–100 per km^2 (2.2 to 5.5 times the overall average) are considerably more extensive. Such areas are virtually absent from Egypt, where there is an abrupt change from very densely to very thinly settled territory at the limits of land irrigated from the Nile, and occur only in a few small pockets in the Sudan. Elsewhere

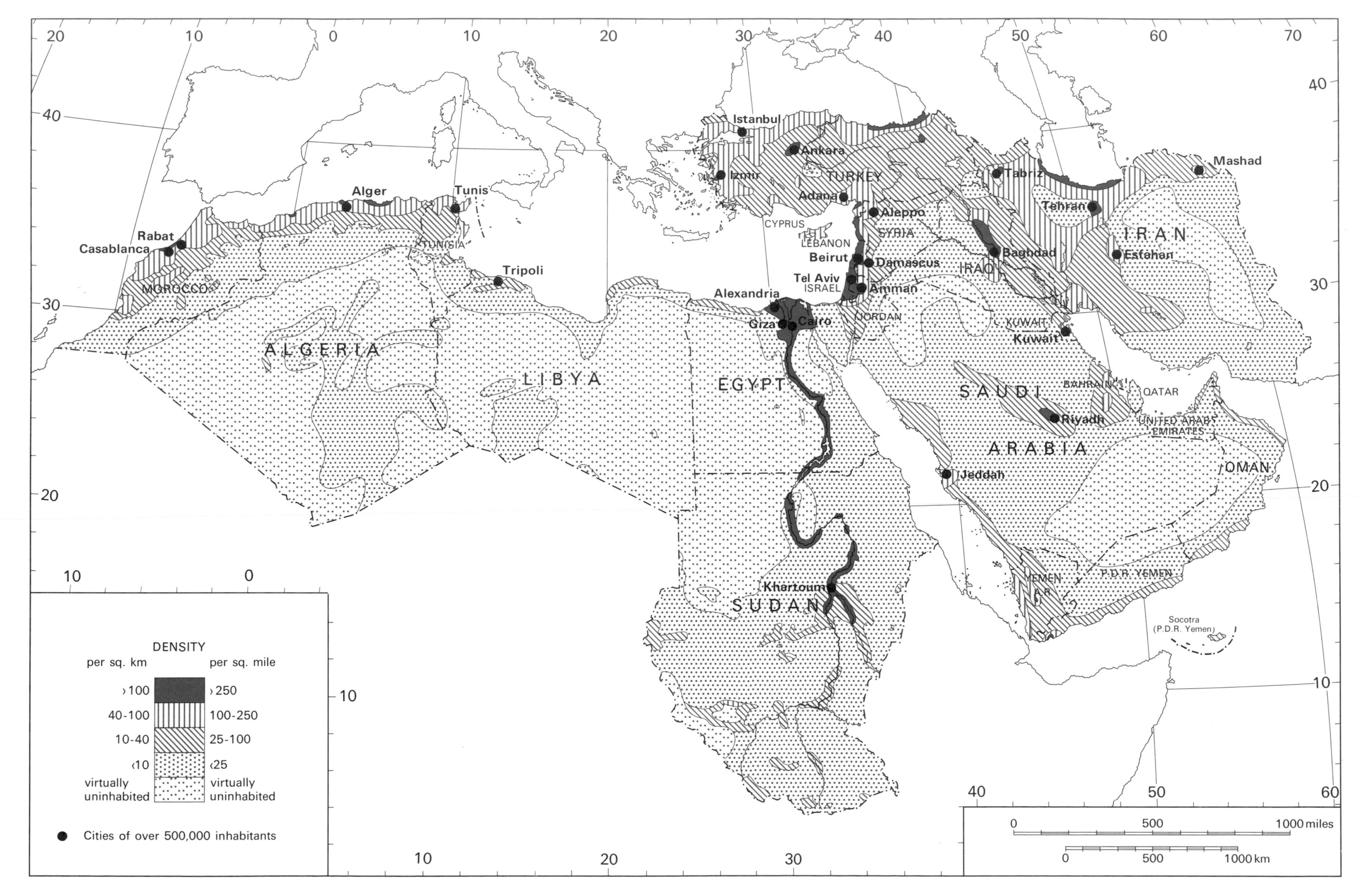

22 Population density

Table 5. *Population by state and region*

	Area		Population		Density	Av. annual
	000 km²	%	millions	%	per km²	% growth 1975–82
North Africa	8,272	55.9	116.8	43.2	14.1	3.2
Maghreb	3,005	20.3	48.1	17.8	16.0	3.3
Algeria	2,382	16.1	20.5	7.6	8.6	3.4
Morocco	459	3.1	20.8	7.7	45.3	3.3
Tunisia	164	1.1	6.8	2.5	41.4	2.5
Northeast Africa	5,267	35.6	68.7	25.4	13.0	3.0
Egypt	1,001	6.8	44.7	16.5	44.6	2.7
Libya	1,760	11.9	3.4	1.3	1.9	4.1
Sudan	2,506	17.0	20.6	7.6	8.2	3.1
Southwest Asia	6,512	44.0	153.5	56.8	23.6	3.0
Northern tier	2,427	16.4	93.7	34.7	38.6	2.6
Iran	1,648	11.1	42.1	15.6	25.5	2.8
Turkey	779	5.3	51.6	19.1	66.2	2.5
Levant/Mesopotamia	759	5.1	34.9	12.9	46.0	3.1
Cyprus	9.3	0.1	0.8	0.3	86.0	0.5
Iraq	438	3.0	14.7	5.4	33.6	3.3
Israel	21	0.1	4.1	1.5	195.2	2.3
Jordan	96	0.6	3.1	1.1	32.3	3.7
Lebanon	10	0.1	2.6	1.0	260.0	0.5
Syria	185	1.3	9.6	3.6	51.9	3.8
Arabian peninsula	3,326	22.5	24.9	9.2	7.5	3.9
Bahrain	0.6	n	0.4	0.1	666.7	3.3
Kuwait	18	0.1	1.7	0.6	94.4	6.4
Oman	272	1.8	1.1	0.4	4.0	3.1
Qatar	11	0.1	0.3	0.1	27.3	6.3
Saudi Arabia	2,400	16.2	10.4	3.8	4.3	4.3
United Arab Emirates	92	0.6	1.2	0.4	13.1	7.0
Yemen A.R.	195	1.3	7.6	2.8	39.0	2.0
Yemen, P.D.R.	337	2.3	2.2	0.8	6.5	3.1
Total	14,7	100.0	270.3	100.0	18.3	3.1

n = negligible
Source: World Bank, *World Development Report 1984* (World Bank/Oxford University Press, New York, 1984).

they cover large areas of relatively productive agricultural territory, as in the main settled zone of the Maghreb, northern and western Turkey, western Iran, eastern parts of the Lebanon and Israel, northwest Syria and the Adana plain of Turkey and parts of Yemen. Rain-fed agriculture is dominant in most of these areas, often supplemented by seasonal irrigation.

Moderate densities of 10–40 per km² (from half to 2.2 times the overall average) are also associated with rain-fed agriculture, though in drier areas the water supply is augmented by irrigation. Such densities are most widespread in the north of the region, where they cover most of the remainder of Turkey, western and northeastern Iran and the 'fertile crescent' zone of eastern and northern Iraq and northern Syria. Densities of 10–40 also occur around the coastal fringes of the Arabian peninsula, in the more southerly parts of the settled zone of the Maghreb, along the Mediterranean coastal fringes of Libya and Egypt and in the main agricultural districts of the southern Sudan.

All these areas of high or moderate density are in marked contrast to the vast areas of thinly settled (less than 10 per km²) or 'virtually uninhabited' territory which are a striking feature of the Middle East and North Africa. The largest continuous area of such territory is in the northern section of the Sahara desert: the greater part of Algeria, Libya, Egypt and the northern Sudan are in the 'virtually uninhabited' category. Semi-desert and savanna areas of the central and southern Sudan have densities mainly below 10 per km². A second major area of thinly settled territory is the interior of the Arabian peninsula, where the southerly Rub al-Khali (the 'empty quarter') and the Nafud desert in the north are virtually uninhabited. Densities are also below 10 per km² throughout the eastern half of Iran and very low indeed in the near-empty Dasht-e-Kavir and Dasht-e-Lut.

It should be realised that the description above, like the map to which it is attached, is highly generalised. Within each of the categories identified, more detailed examination at a larger scale would reveal a still greater range of variation in population density. Such local variations are associated not only with the physical environment and the nature of the local agriculture but also with the presence or absence of sizeable towns. Only towns with 500,000 or more inhabitants are shown on Map 22 – hence the high density pockets around Ankara, Tehran, Tabriz, and so on; a much larger number of towns is shown on a later map (Map 26).

Population change

Population data from most of the countries of the Middle East and North Africa are somewhat inadequate as a source of information for the calculation of population change. Few of the 22 countries in the region have an established system of vital registration recording all births and deaths and thus providing an accurate measure of their rates of natural increase, the data on international migration are fragmentary and incomplete and in several cases the absence of a reliable run of censuses presents difficulties in calculating even the net intercensal changes which have occurred. Thus the rates of growth at regional and national level given in Table 5 and those mapped at the country level (Map 23) are informed estimates rather than completely accurate figures.

There can be no doubt that, over the region as a whole, population growth has continued at a rapid rate in recent years. In practically every country, crude death rates have fallen to between 10 and 15 per thousand (with rates well below 10 in several cases), whereas crude birth rates have remained at high levels, generally between 35 and 45 per 1000, producing natural increase rates between 2.5% and 3.5% per annum during the late 1970s and early 1980s.

Relatively few countries have growth rates outside the 2.5–3.5% range. Where they are significantly higher, this usually reflects the impact of immigration into oil-producing countries with an inadequate indigenous labour force, as in Kuwait, Qatar, the United Arab Emirates and, to a lesser extent, Saudi Arabia and Libya. Growth rates significantly below average occur in some instances as a result of the persistence of rather high death rates (e.g. Yemen A.R.) and in others as a result of significant fertility decline (e.g. Cyprus, Israel). At current growth rates, the population of the Middle East and North Africa is increasing by 7 or 8 million a year and is likely to exceed 400 million by the end of the century. Given the very youthful age structure of all the populations of the region, with 40–45% below the age of 15 and only 2–4% above the age of 64, and the slow progress so far achieved in the reduction of fertility, rapid growth is set to continue through the 1980s and 1990s.

The likely effects of continuing rapid population growth are indicated by recent population projections which suggest that by A.D. 2000 the Middle East and North Africa will have a total population in the vicinity of 420 million. This represents an increase of some

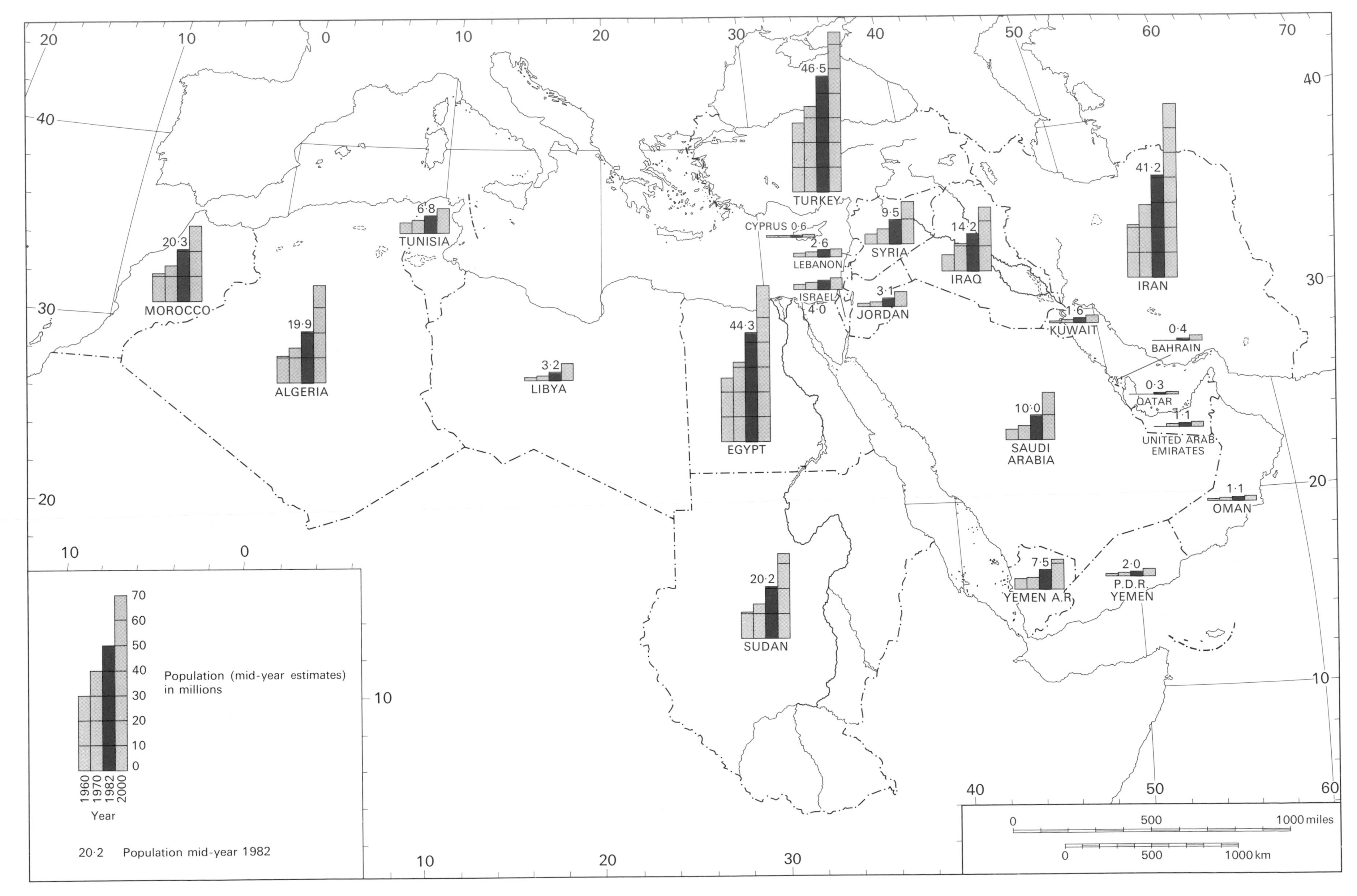

23 Population change

55% from the total of the early 1980s (see Table 5). Projected growth rates vary widely on either side of this average figure; only two countries – Israel and Lebanon – are likely to experience an increase of less than 25% during the final two decades of the twentieth century, and several are predicted to experience more than 75% growth, notably Algeria, Libya, Iraq, Jordan, Syria and Saudi Arabia. There will be some striking changes in the absolute sizes of national populations. By A.D. 2000, Egypt, Iran and Turkey will each have more than 60 million inhabitants. Algeria, Morocco and Sudan will have more than 30 million, Iraq 26 million, Saudi Arabia 19 million and Syria 17 million.

Living standards

The population growth rates discussed in the previous section are, of course, a product of the birth, death and resultant natural increase rates; except in the case of a few of the smaller countries, most notably the Gulf states, migration is only a minor factor. Given that, in most countries of the region, fertility is likely to decline only slowly, levels of mortality will be a major influence on rates of population increase for the remainder of the century. Certainly the countries which are now experiencing the most rapid population growth are those which have been most successful over recent years in reducing their mortality levels. Particularly important in this respect is the level of infant mortality, which is also considered to be an important indicator of level of development and standard of living. Twenty years ago, infant mortality was high in virtually every country. It exceeded 100 (infant deaths per 1000 live births) in 20 of the 22 countries (the two exceptions being Israel and Lebanon), and was above 150 in half of them. Over the past two decades, every country in the region has achieved a considerable reduction in the level of infant mortality, which has been halved in many cases. Despite this welcome trend, rates in most countries remain well above those of the developed world (Israel is, once again, an exception with a rate of only 16 infant deaths per 1000 live births) and there is scope for further improvement. Rates above 100 are still recorded in nine countries; the highest rates (125 or more) are in Sudan and the Yemens, where one child in eight is likely to die before its first birthday.

Such variations are, of course, related mainly to the level of medical services, one measure of which is the number of doctors per 10,000 population. This has risen significantly in every country, but wide discrepancies remain. The oil-rich states now have relatively large numbers of doctors: 18 per 10,000 population in Kuwait, for example (compared with 15 in the U.K. and 24 in the U.S.A.), 14 in Libya and 11 in the United Arab Emirates, several countries with a long tradition of medical education also have quite high figures. 19 in Lebanon, 10 in Egypt, 6 in Turkey. At the other extreme, there is only one doctor for 5000 or more people in Morocco, Sudan and the Yemens.

Standards of medical care and reductions in the level of mortality are reflected in figures for the average expectation of life at birth. In 1960, with the exception only of Israel and Lebanon, life expectancy for males throughout the region was less than 50 years and in some countries – the Gulf states, the Yemens and Sudan – it was less than 40 years. Since then it has risen in every case. Except in Sudan (46 years) and the Yemens (45 years), it is everywhere above 50 and exceeds 60 in Iran, Israel, Jordan, Kuwait, Lebanon, Syria, Tunisia and the Gulf states. In every country, life expectancy for women is three or four years greater than that for men.

Such data indicate significant improvements in standards of living throughout the region. Further reductions in mortality are likely and desirable, but will inevitably result in continuing rapid population growth.

KEY REFERENCES

Clarke, J. I. and Fisher, W. B. (eds.), *Populations of the Middle East and North Africa* (University of London Press, London, 1972).

Hill, A. G., 'Population growth since 1945' in J. I. Clarke and H. Bowen-Jones (eds.), *Change and Development in the Middle East* (Methuen, London, 1981).

World Bank, *World Development Report 1984* (World Bank/Oxford University Press, New York, 1984).

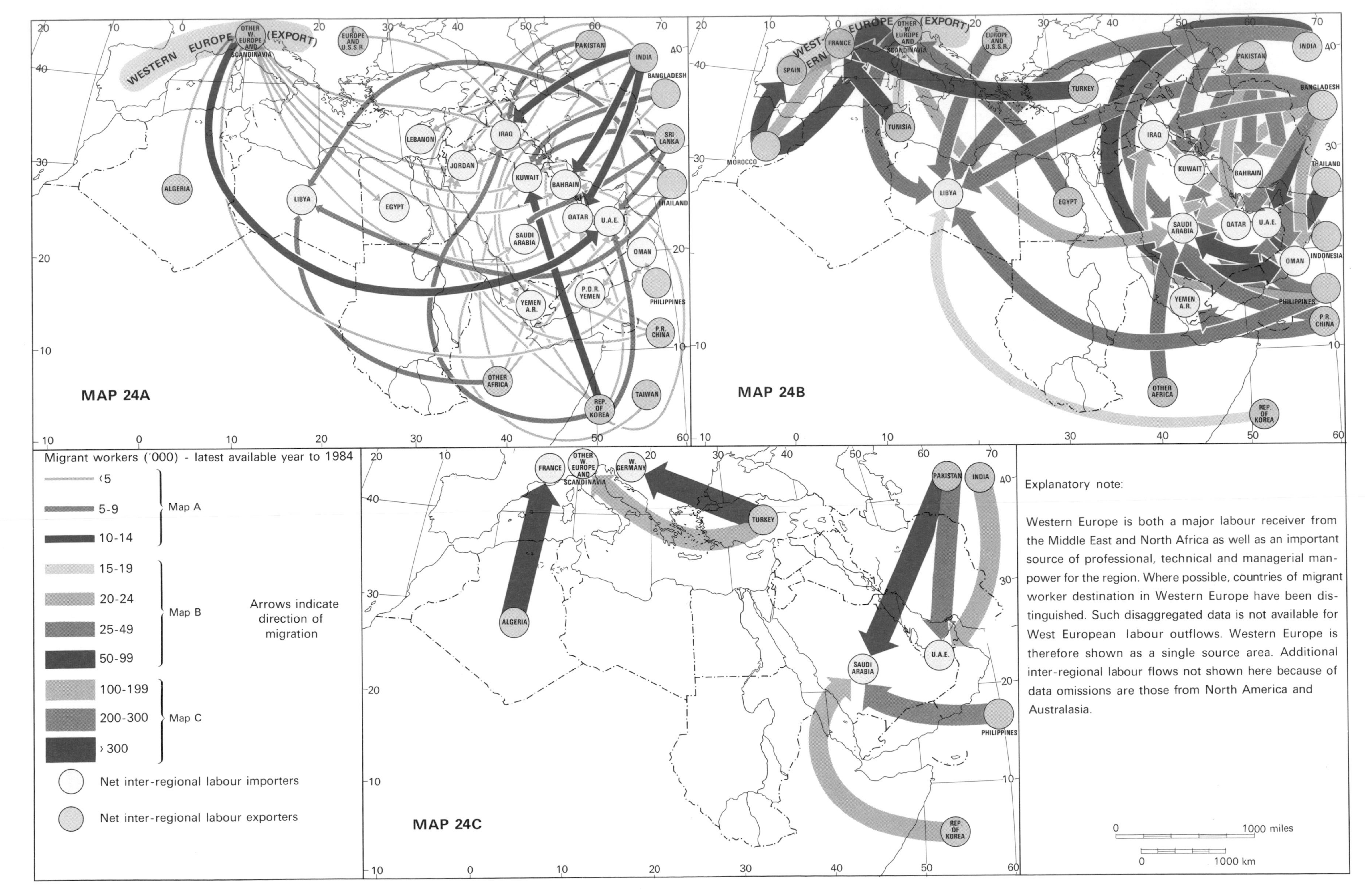

24 Inter-regional labour migration

Inter- and intra-regional labour migration

Ian Seccombe

International labour migration has had a significant and sometimes dramatic effect on social and economic structures in almost all the countries of the Middle East and North Africa. Today, more than 5.5 million migrant workers are employed outside their home countries. Until the early 1970s labour emigration was dominated by labour outflows to Western Europe from Turkey and the three Maghreb states (Algeria, Tunisia and Morocco). The latter movement was primarily, though not exclusively, to France and, particularly in the case of Algeria, had begun during the French colonial period. With independence the three North African governments adopted pro-emigration policies as a solution to their growing problems of unemployment and underemployment. Although bilateral agreements were signed with several European countries, France remained the preferred destination for most emigrant workers. By 1973 the North African community (workers and their dependents) in France numbered more than 1.2 million. The majority of Maghrebin migrants were engaged in low-paid, unskilled and often arduous or menial tasks in the services, manufacturing and construction sectors. Similar jobs, increasingly unacceptable to European nationals, were taken up by the growing number of Turks absorbed by West Germany's post-war economic reconstruction. Between 1963 and 1973 the number of Turkish *Gastarbeiter* in the Federal Republic grew from 22,000 to 615,000, accounting for over 78% of Turks working abroad.

Western Europe's economic recession, dating from 1973, led to a sudden and marked shift in immigration policies. In November 1973 West Germany imposed a ban on the entry of new, non-European Community labour inflows. By the end of the following year most West European countries had instituted similar restrictive policies and were, in the face of rising unemployment, seeking to encourage immigrant workers to return home. However, large-scale return migration has not occurred and, despite the ban on new labour inflows, the Turkish and Maghrebin communities in Western Europe have continued to grow through a combination of natural increase, family regroupment and clandestine immigration. By 1982 more than 3.5 million Turks and North Africans were living in Western Europe, of whom 1.5 million were economically active.

The oil price increases of 1973–4 had a dramatic effect on the scale and pattern of international labour migration in the Middle East and North Africa. While employment opportunities in Western Europe were declining, new demands for expatriate manpower were stimulated within the region itself. Tunisian emigration, for example, turned increasingly to Libya. Increased oil revenues accruing to the main oil-exporting countries (Saudi Arabia, Libya, Kuwait, the United Arab Emirates, Qatar, Bahrain, Oman, Iraq) effectively removed financial constraints to economic growth. By 1975 these states had a foreign labour force of about 1.6 million, representing some 17% of their total workforce, compared to less than 800,000 in 1971–2. Although expatriate labour, particularly in technical and professional occupations, had been employed in the Gulf states since the first commercial exploitation of oil resources in the mid-1930s, the scale of labour inflows in the 1970s was unprecedented.

The major factor behind the growth in immigration into and within the Arab region is the small size of the indigenous labour force, and its limited range of skills. The combined effects of small population size, youthful age structure and social constraints on women's participation in the economy have resulted in low labour force participation rates. As a result, accelerated economic growth, in particular in the large-scale investment in infrastructural construction projects and industrial development plans, inevitably generated manpower demands far in excess of indigenous supply capacity. By 1980 the immigrant labour force had reached over 3 million as countries such as Oman and Iraq, formerly labour exporters, also became significant labour receivers. Recent estimates suggest that in 1985 the immigrant labour force in the Arab world may have exceeded 4.3 million workers, representing some 28% of the total workforce.

The greatest number of foreign workers are employed in Saudi Arabia and Libya which together in 1980 accounted for over half of all migrant workers in the region. However, the relative dependence on non-nationals in the labour force is greatest in the smaller Gulf states such as Kuwait, where immigrants account for 66% of the total workforce, Qatar (86%) and the United Arab Emirates (90%). Even in Saudi Arabia the share of total employment accounted for by non-nationals has grown from 34% in the mid-1970s to over 50% in 1985.

In the mid-1970s the majority (65%) of these migrant workers came from surrounding Arab states, in particular Egypt, which accounted for 22% of migrant workers in the region, the Yemen Arab Republic (20%) and Jordan (14%). The latter figure includes many Palestinians who emigrated to the Gulf states after the June 1967 Arab–Israeli war. South Asians, particularly from India and Pakistan, were also an important source of manpower (22% of the total) especially in the lower Gulf states (Bahrain, Qatar, U.A.E. and Oman) with which they had long-standing links.

Unlike the Maghrebin and Turkish workers in Western Europe, immigrants in the Arab region are employed at all occupational skill levels, from professional and technical staff (of which less than 20% is indigenous) to unskilled labourers. A significant proportion of professional and managerial workers are drawn from the United States and Western Europe; in 1983 almost 15,000 British workers left the United Kingdom for employment in the Middle East. Overall, the largest proportion of migrant workers (56% of the total) are employed in the construction and service sectors. During the 1980s structural developments in the economies of the capital-rich states have increased the demand for skilled and professional workers while the demand for unskilled migrants is falling with the completion of many major construction programmes. The disproportionate and increasing loss of high-level manpower has become a major disadvantage for the Arab labour-supplying states through the aggravation of existing skill scarcities. Inflexibility within the labour market and education system has ensured that even small withdrawals of skilled manpower can have a significant impact on production. The emergence of such skill shortages in a number of labour-supplying states has stimulated domestic wage inflation. In response a second, compensating, flow of non-national labour into such labour-sending economies has occurred. Such shortages were particularly prevalent in Jordan which by 1975 had an estimate 40% of its manpower working abroad. By 1982 Jordan employed more than 130,000 immigrant workers (predominantly Egyptians) to 'replace' those Jordanians working abroad.

Arab labour-supplying countries which originally welcomed labour outflows as a means of reducing high rates of underemployment and unemployment and of increasing foreign exchange earnings have become more concerned with the negative consequences of their participation in the international labour market. The selectivity of manpower withdrawals and in particular the loss of skilled manpower has reduced the rate of growth in the domestic economy and in some cases may have negated the potentially beneficial impact of emigration on unemployment. At the same time labour emigration has had a sometimes disruptive effect on the traditional sectors of labour-supplying economies, leading to falling agricultural production with the collapse of farm terraces in Yemen A.R. and the deterioration of *falajes* in Oman. In Yemen A.R. there has been a switch from export crops like coffee and cotton to qat (a popular local stimulant), which has a much lower labour input but cannot be exported. A number of recent studies have demonstrated growing social and economic division within areas of emigration, based on the relative success (or failure) of emigrants vis-à-vis their non-migrant neighbours.

Divergence between the social and private returns to labour migration are most readily apparent in the growing disenchantment expressed by Arab labour-suppliers with the contribution of workers'

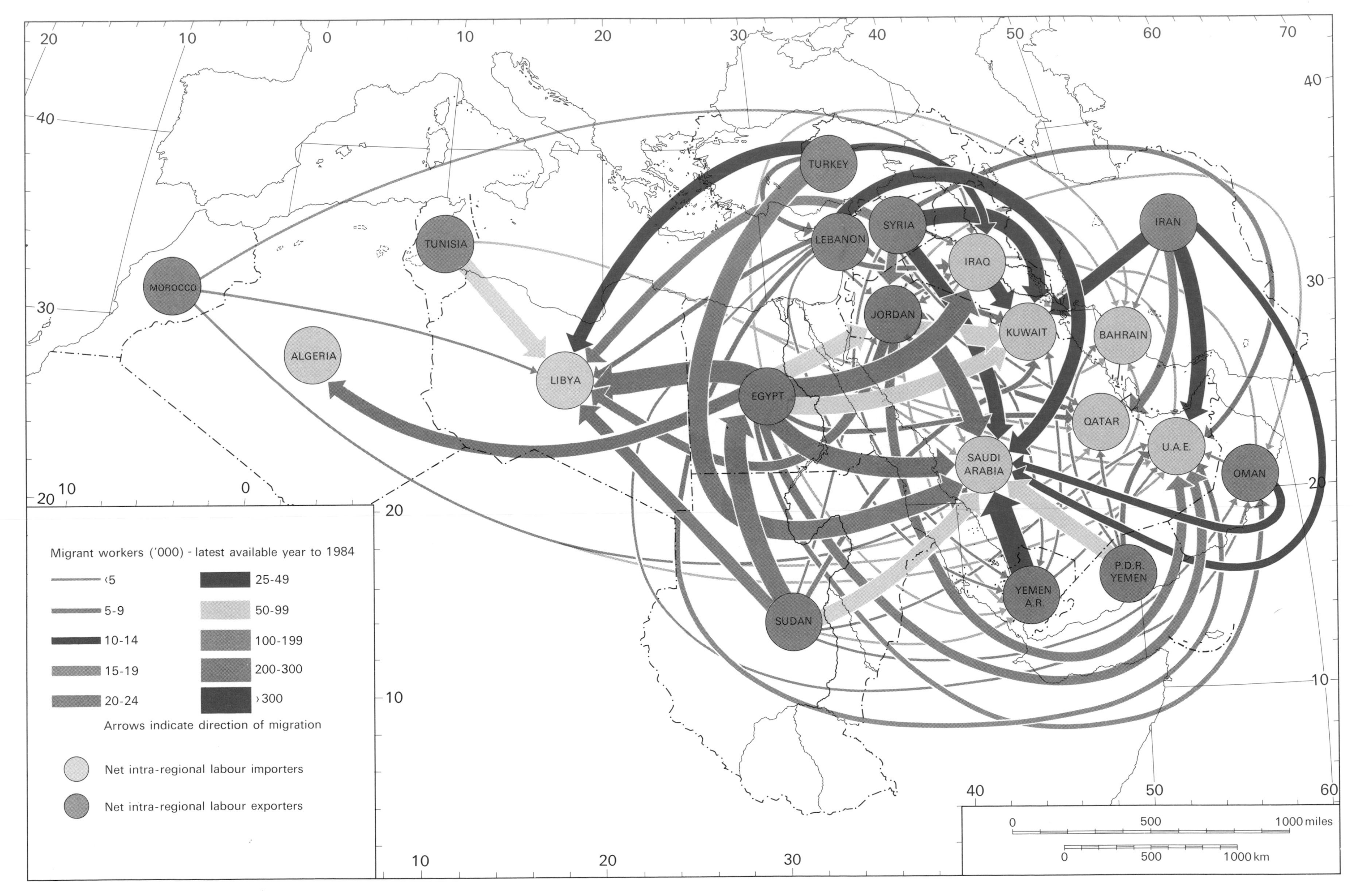

25 Intra-regional labour migration

remittances to economic growth and development. Although remittance earnings have been welcomed as a source of foreign exchange to offset trade imbalances (in Yemen A.R., for example, remittances in 1982 stood at 4000 million Yemeni rials compared to domestic exports of only 22 million Yemeni rials), their potential as a source of savings and investment capital has been far from realised. Available evidence confirms that the propensity to consume remittance income is high; migrants' purchases of land, housing and consumer durables (often with a high import content) has contributed to inflation without adding to the productive base of the economy. While remittance expenditure clearly benefits the consuming household (in the short term, at least), there are inherent dangers at the national level in the reliance on an uncertain income source. The level of remittance receipt depends not only on employment and exchange policies in both the labour-supplying and receiving countries, they are also influenced by the demographic evolution of the migrant communities themselves.

The growth in the expatriate workforce in the labour-receiving states has been accompanied by the development of significant immigrant communities with the settlement of migrant workers' dependants in the host countries. In 1975 the estimated 1.6 million migrant workers in the region were accompanied by 1.7 million dependants, and in the smaller Gulf states non-nationals were already a majority of the total population. By 1985 the number of dependants probably exceeded 6 million. The demographic evolution of these immigrant communities poses a number of crucial policy issues for the labour-receiving countries. As well as adding to the costs of infrastructural and service provision, immigrant communities, which in a number of cases outnumber the indigenous population, are increasingly seen to pose a threat to national culture and identity. Rising political aspirations among expatriate communities may pose a threat to the political stability of their largely conservative host regimes. Future capital investment decisions may be influenced by the need to compromise between the demands of economic diversification (based on further immigrant labour flows) and the requirements of internal stability. This conflict is seen by some as an important element in recent developments in the organisation and patterns of migrant labour flows in the region.

Since 1975 the pattern of migrant origins has changed significantly, the Arab share of labour inflows falling from 65% in 1975 to 48% in 1985. The stock of Asian migrant workers in the Middle East is currently estimated at between 1.3 and 2.6 million. The initial boom in Asian labour flows came from South Asia, particularly Pakistan and India. In the later 1970s increasing numbers of southeast and east Asians entered the regional labour market and by 1980 accounted for 40% of all Asian workers in the region. Particularly prominent labour suppliers are the Republic of Korea, the Philippines and Thailand. More recently these have been joined by Indonesia and the People's Republic of China. Turkey has also become an important source of immigrant labour in the Arab world (particularly in Libya, which is also a major importer of East European manpower) following the closure of the Western European labour market to new inflows.

Asian labour migration to the Middle East has been characterised by a high degree of organisation involving private and state-run recruiting agencies able to identify and meet specific employer requirements. In addition the growth of Asian immigration and employment has been closely associated with the penetration of national contracting companies into the Middle East construction market on turn-key and enclave projects. Dependants are prohibited from accompanying the migrant workers who are under the control of their employers throughout the contract period. The company provides food, accommodation, medical services and recreational facilities on a work camp basis. On the completion of the contract the workers are repatriated by the company. Thus while providing a more cost-effective labour force the organised supply of manpower also minimises the social and political implications associated with hosting a large Arab immigrant workforce and community.

Current economic conditions in the Gulf states suggest that recent projections, which predicted the net addition of about 1.7 million foreign workers by 1990, are far too high. The downturn in world oil prices since 1981 has considerably reduced government revenues. Indeed Saudi Arabia has reduced its 1985–90 development budget by 20%, while one of the key objectives of the new development plan is the reduction, by 600,000, of the foreign workforce. Similar proposals have been made in Kuwait, where the government plans to balance the national and expatriate components of the population by the year 2000.

Aside from planned reductions in expatriate employment, there is growing evidence of high rates of worker emigration. The first half of 1985 saw up to 60,000 foreign workers leaving Saudi Arabia each month, while in the U.A.E. the departure of expatriates led to a 2% fall in the population total in 1984. In August and September 1985 Libya pursued a policy of forced expulsions of thousands of expatriate workers from Egypt, Mali, Mauritania, Niger, Senegal and Tunisia. The deportations were directed at nationals of countries with whom Libya has poor international relations, particularly Tunisia. More than 29,000 Tunisians were expelled. Although the 1985 wave of deportations from Libya had political overtones similar to previous expulsions in 1976 and 1980, they were also symptomatic of a wider fall in the regional demand for foreign labour.

The current recession clearly offers the labour-importing states a breathing space in which to re-evaluate policies towards the employment of both domestic and immigrant labour, since it is only through the promotion of more active participation by the indigenous population that the level of foreign worker dependence can be reduced in the long term.

Falling demand for foreign labour presents the labour-sending countries with a dual challenge. Naturally, remittance earnings, which are the major source of foreign reserve in Egypt, Jordan and the Yemens, are likely to fall significantly with the return of unskilled and semi-skilled migrants, thus forcing further cuts in government expenditure. At the same time there is likely to be a dramatic growth in unemployment. To date, however, none of the Arab labour-sending countries have drawn up contingency plans to cope with the likely return and reintegration of their migrant workforce. Clearly, international labour migration has become and will remain a major influence upon economic development in the Middle East.

KEY REFERENCES

Al-Moosa, A. R. and McLachlan, K., *Immigrant Labour in Kuwait* (Croom Helm, London, 1985).

Arnold, F. and Shah, N. M., 'Asian labour migration to the Middle East', *International Migration Review*, 18 (2), 1984, pp. 294–318.

Birks, J. S. and Sinclair, C. A., *Arab Manpower: The Crisis of Development* (Croom Helm, London, 1980).

Owen, R., *Migrant Workers in the Gulf*, Minority Rights Group Report no. 68 (M.R.G., London, 1985).

Serageldin, I. *et al.*, *Manpower and International Labour Migration in the Middle East and North Africa* (Oxford University Press, Oxford, 1983).

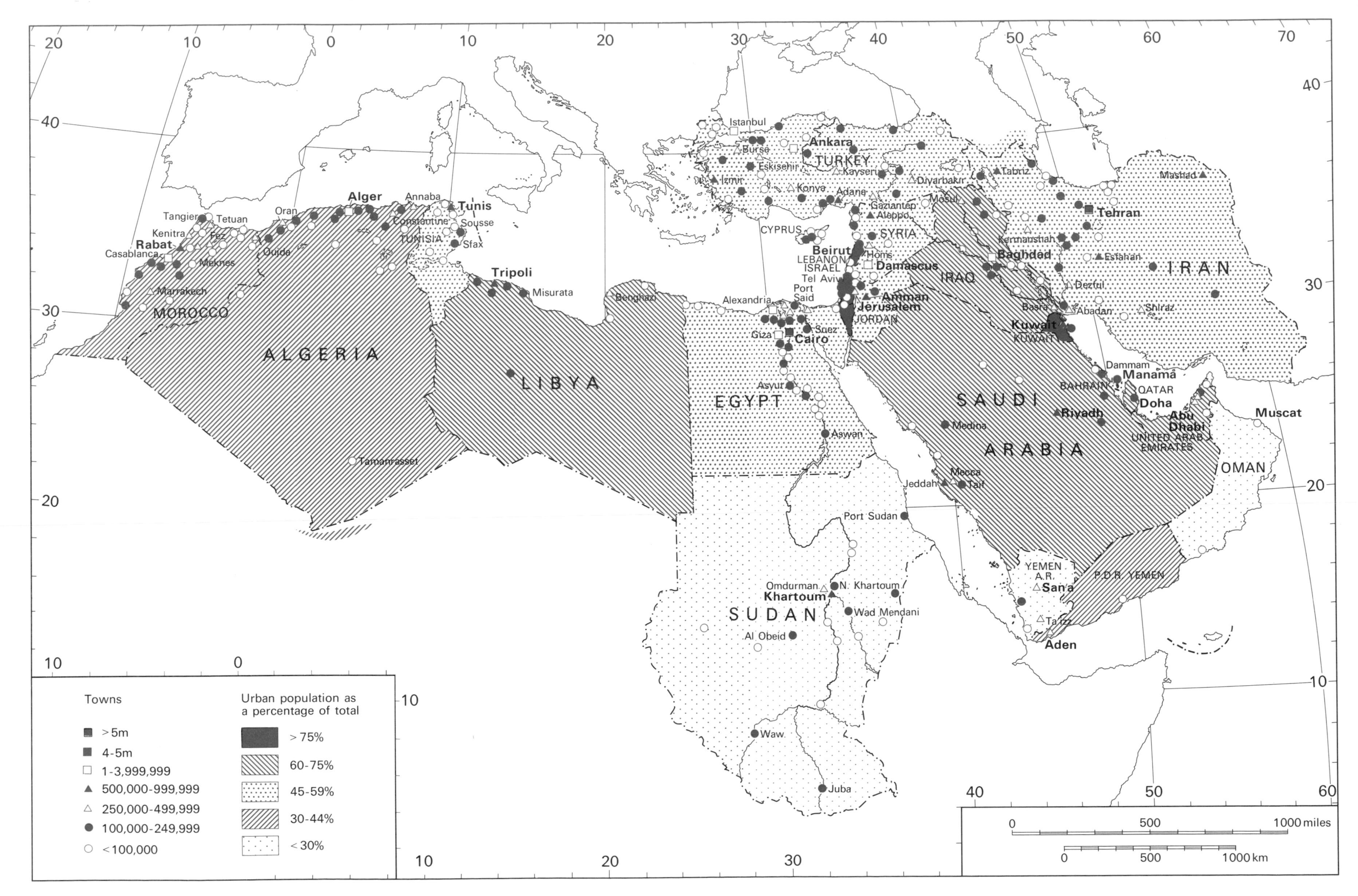

26 Urban populations and towns

Urban population and towns

John Dewdney

The region has a long tradition of urban life and saw the emergence of some of the world's earliest cities at least 5000 years ago in Mesopotamia and the Nile valley. In recent decades, the countries of the Middle East and North Africa have experienced rapid growth both in the numbers of people living in towns and in the size of many individual cities. Problems arise, however, in expressing these trends numerically and particularly in making comparisons of the levels of urbanisation in the various countries and in attempting to rank the cities by size. All the countries of the region produce statistics which purport to show the size of their urban populations and of individual towns, but there is no common definition of the term 'urban' which can be applied throughout the region. In Turkey, for example, the urban population is defined as the population living within the administrative centres of provinces and districts; in Israel it is the population of all settlements with more than 2000 inhabitants (excluding those where one-third or more of the gainfully occupied heads of household are employed in agriculture); in Iran it is the population of all *shahrestan* (province) centres plus other places with 5000 or more inhabitants. It is clear that the population of a settlement of a particular size or administrative status may be classed as urban in one country but not in another, so that international comparisons are not, strictly speaking, valid in many cases. Thus the urban proportions displayed in Map 26 and the numbers and proportions of town-dwellers listed in Table 6 show only the urban populations as officially defined in each country, a point which it is important to bear in mind in connection with the discussion which follows.

Of the total population of the region, 123 million or 50% are classed as urban, the proportion being significantly higher in Southwest Asia (54%) than in North Africa (45%). While the level of urbanisation is still well behind that of the advanced industrial nations, it is quite high in comparison with other developing countries (e.g. China 25%; India 22%). A wider range of values is apparent at the country level. Those with the larger populations and a considerable agricultural potential are generally within the range of 40–55% urban (e.g. Egypt, Iran, Turkey, Algeria and Morocco); a notable exception is the case of Sudan which, with only 21% urban, is more akin to the countries of tropical Africa than to those of the Middle East proper. Higher levels of urbanisation reflect a variety of special conditions. Israel (88% urban, but note the low threshold value mentioned above) is a unique case in this as in so many other respects; Lebanon (80%) is noteworthy for the importance of commercial activities. Very high levels of urbanisation are also recorded in some of the Gulf states (Bahrain, Kuwait, Qatar, U.A.E.) and in Saudi Arabia, all countries where the agriculture potential is very limited. At the other extreme, the level is very low in Oman and Yemen A.R.

While many of the centres officially defined as urban for census purposes are small and have few urban functions, there can be little doubt concerning the truly urban character of places with more than 100,000 inhabitants. Data for these are also provided in Table 6, though here, too, care is necessary: the methods used to identify and delimit urban agglomerations are not the same in every country. Roughly a quarter of the total population and about half of those classed as urban live in such centres, of which there are now nearly 170, compared with about 80 in the 1960s.

Table 6. *Urban populations*

	Urban population		Living in towns of 100,000+			
	millions	%	No. of towns	Pop. (millions)	% of urban	% of total
North Africa	48.5	45	67	27.8	57	26
Maghreb	21.8	47	33	11.6	53	24
Algeria	9.1	44	15	4.2	46	21
Morocco	9.1	44	15	5.7	63	27
Tunisia	3.6	53	3	1.7	47	25
Northeast Africa	26.7	42	34	16.2	61	26
Egypt	20.1	45	21	12.1	60	27
Libya	2.2	60	4	2.0	90	55
Sudan	4.4	21	9	2.1	48	10
Southwest Asia	74.8	54	98	38.7	52	28
Northern tier	41.3	47	51	21.1	51	23
Iran	20.9	50	22	9.8	47	23
Turkey	20.4	44	29	11.3	55	23
Levant/Mesopotamia	22.2	63	28	10.9	49	31
Cyprus	0.3	40	2	0.3	98	39
Iraq	9.8	67	8	4.4	45	30
Israel	3.6	88	8	1.6	44	39
Jordan	1.7	55	3	0.9	53	29
Lebanon	2.1	80	2	0.9	43	34
Syria	4.7	49	5	2.8	60	29
Arabian peninsula	11.3	53	19	6.7	59	31
Bahrain	0.3	72	1	0.1	33	25
Kuwait	1.4	86	3	1.0	71	63
Oman	0.2	18	0	0	0	0
Qatar	0.2	67	1	0.1	50	33
Saudi Arabia	6.7	64	7	3.5	52	35
U.A.E.	0.8	67	3	0.8	100	73
Yemen A.R.	0.9	12	3	0.9	100	12
Yemen, P.D.R.	0.8	37	1	0.3	37	15
Total	123.3	50	165	66.5	54	27

Source: Based on data for *c.* 1982. W. M. Willett, *The Geographical Digest 1984* and *The New Geographical Digest* (George Philip, London, 1984 and 1986).

Map 26 shows urban centres listed in the various national census reports and other sources, classified into groups according to population size. The general distribution of towns is clearly related to the maps of population distribution and density (Maps 21 and 22), with pronounced concentrations in the Nile valley and along the Levant coast, a broader scatter in areas where the population as a whole is more evenly distributed (the Maghreb, Turkey, western Iran) and widely spaced towns in Sudan, the Arabian peninsula, Syria and Iraq. The importance and rapid growth of capital cities and ports is a feature throughout the region. Of the larger cities (over 100,000) rather more than half are to be found in 5 of the 22 countries: 29 in Turkey, 22 in Iran, 20 in Egypt, and 15 in Algeria and in Morocco.

Until quite recently, the majority of the countries showed a high level of urban primacy, with the largest centre – usually the capital – standing head and shoulders above all its rivals. This still applies in several cases: the population of Tehran, for example, at 5.7 million, is six times that of the next largest city, Esfahan (672,000); Baghdad at 3.2 million is eight times the size of Basra – but this situation is becoming less common, and there are several cases of 'double primacy': Istanbul (2.8 million) and Ankara (1.9 million) in Turkey; Damascus (1.2 million) and Aleppo (920,000) in Syria; Riyadh (667,000) and Jeddah (561,000) in Saudi Arabia; Jerusalem (407,000) and Tel Aviv–Yaffo (335,000) in Israel, for example. Several countries, including Turkey, Morocco, Algeria and, to a lesser degree, Iran, now have networks of sizeable urban centres. The situation in Egypt is worthy of special mention. Although the level of urbanisation in that country, at 45%, is close to the regional average, the concentration of the urban population into large cities is particularly striking. From Cairo to the Mediterranean there are a dozen towns with more than 100,000 inhabitants each, including three of the region's nine 'millionaire' cities – Cairo (10 million, the largest city in the Middle East), Alexandria (2.3 million) and Giza (1.2 million). Egypt, Turkey and Iran together contain half of all the town-dwellers in the Middle East and North Africa.

Throughout the region, as we have already seen (Maps 21 and 22), the distribution of population is extremely uneven. The rapid growth of cities is increasing this unevenness and leading to an ever-greater concentration of population in the more urbanised areas. There can be no doubt that, by the end of the present century, town-dwellers will outnumber the rural population, a situation already reached in several of the smaller countries.

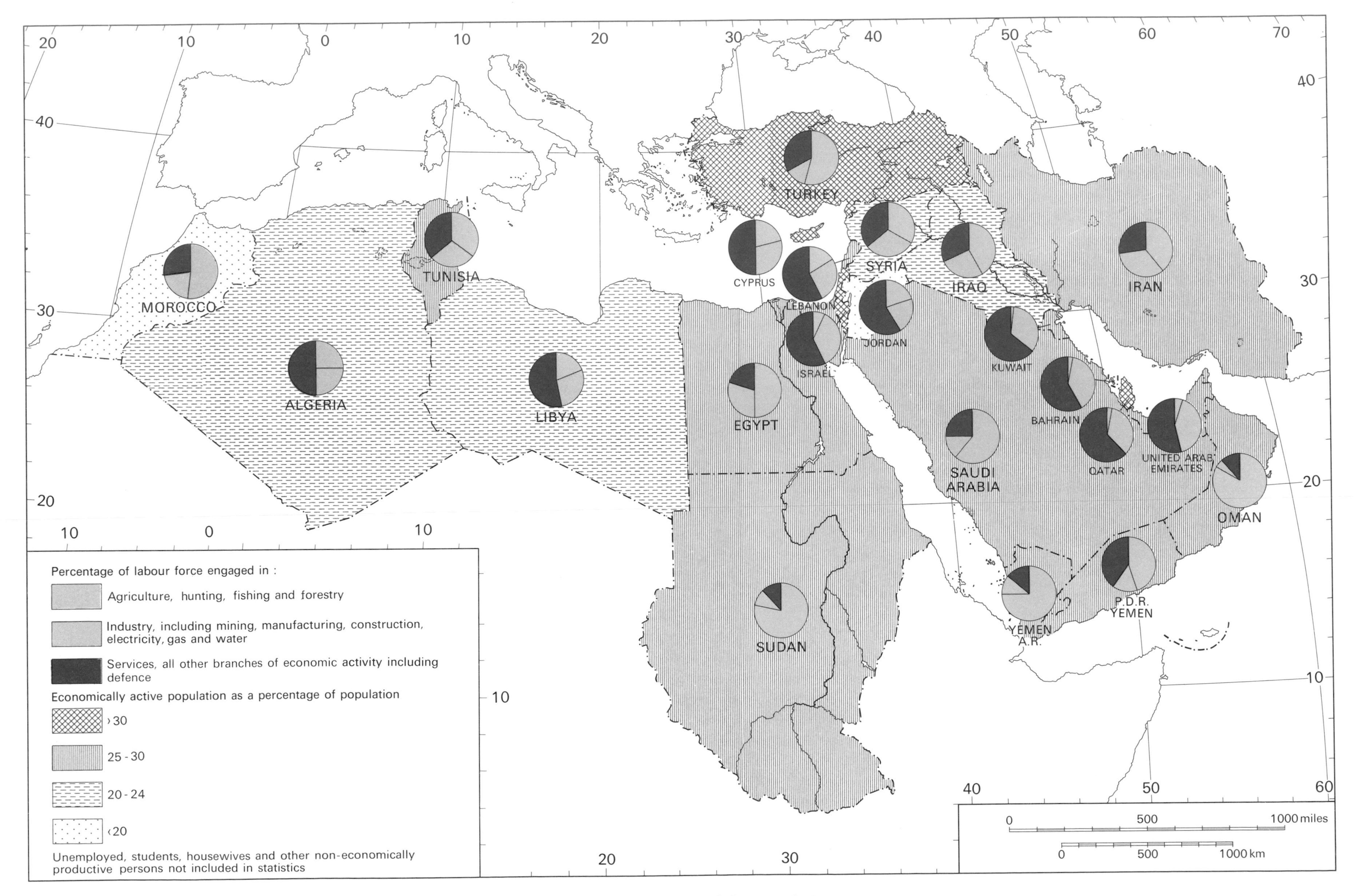

27 Economic structure of the population

MAP 27

Population: economic structure

John Dewdney

Map 27 shows two aspects of the labour force of the region: the proportion of the total population defined as 'economically active' is indicated by shading; the distribution of the economically active among the three major sectors of the economy – agriculture, industry and services – is depicted by divided circles (pie diagrams). As with many of the other elements portrayed in this atlas which are based on national statistics, international comparisons present problems owing to the different definitions of economically active and of the three sectors used by individual countries.

The proportion of any national population which is economically active depends on a variety of factors, demographic, economic and social. In the first place, the size of the potential labour force depends on the population's age structure, which determines the number of people of 'working age'. There is no universally applied definition of this category; Table 7 indicates the proportions aged 15–64, the age range from which the bulk of the labour force is normally drawn. It will be observed that most of the countries of the Middle East and North Africa have 45–55% of their populations within this category, a level much lower than those of the developed industrial nations (U.K. 63%, U.S.A. 67%). This contrast results from the very youthful age structures of the population of the region, which in turn result from high fertility. The extent to which those of working age are actually employed is affected both by the state of the economy – which influences the levels of unemployment – and by such socio-cultural matters as attitudes towards the employment of women. In most of the countries of the region, the economically active constitute between 20% and 30% of the total population (compared to about 40% in the U.K.). Values above 30% occur mainly in countries where the proportion of 16–64-year-olds is above the regional average (e.g. Cyprus, Israel, Lebanon, Turkey) and the lowest values in countries with particularly youthful age structures (e.g. Iraq, Libya, Syria).

Table 7 shows both the proportions of the economically active population employed in the three main sectors and the changes which occurred in those proportions between 1960 and 1980. Given the different classification systems used in the various countries, the most important distinction is that between agricultural and non-agricultural employment. It is readily apparent that, in terms of the employment which it provides, agriculture remains a major element in the economies of most of the countries of the region, particularly those with the larger populations: 40% or more of the economically active are still employed in the agricultural sector in Morocco, Egypt, Sudan, Iran, Turkey, Iraq, Saudi Arabia, Oman and the Yemens, countries which together contain nearly 80% of the region's population. At the other end of the scale, agricultural employment is particularly low in countries with well-developed commercial activities, such as Israel and Lebanon, and in oil-rich states with very limited agricultural potential such as Libya, Bahrain, Kuwait and the United Arab Emirates. It is equally clear that, over these two decades, all the countries of the region experienced a significant decline in the proportion of their labour force employed in the agricultural sector, though in most cases, since the labour force has grown rapidly, the numbers so employed have actually increased. This decline in the relative importance of agricultural employment has been particularly striking in the case of Libya, whose economy has been transformed, and least apparent in Sudan, Saudi Arabia and Yemen A.R., where agricultural employment remains dominant to a very pronounced degree.

Table 7. *Composition of the labour force*

	Aged 15–64 (%)	Percentage employment of the economically active					
		Agriculture		Industry		Services	
		1960	1980	1960	1980	1960	1980
Algeria	49	67	25	12	25	21	50
Morocco	52	62	52	14	21	24	27
Tunisia	52	56	35	18	32	26	33
Egypt	56	58	50	12	30	30	20
Libya	47	53	19	17	28	30	53
Sudan	53	86	78	6	10	8	12
Iran	52	54	40	23	34	23	26
Turkey	56	79	54	11	13	10	33
Cyprus	65		21		29		50
Iraq	48	53	42	18	26	29	32
Israel	59	14	7	35	36	51	57
Jordan	49	44	20	26	20	30	60
Lebanon	52	38	11	23	27	39	62
Syria	47	54	33	19	31	27	36
Bahrain	53		4		36		60
Kuwait	54	1	2	34	34	65	64
Oman	52		80		4		16
Qatar	52		3		37		60
Saudi Arabia	52	71	61	10	14	19	25
U.A.E.	63		5		41		54
Yemen A.R.	49	83	75	7	11	10	14
Yemen, P.D.R.	49	70	45	15	15	15	40

Note: Blanks indicate information not available.
Sources: Age – Population Reference Bureau, *World Population Data Sheet* (Population Reference Bureau, New York, 1983).
Activity – World Bank, *World Development Report 1984* (World Bank/Oxford University Press, New York, 1984).

The allocation of non-agricultural employment to the industrial and service sectors is strongly influenced by national classifications of economic activity. According to the data in Table 7, the majority of the countries of the Middle East and North Africa now have 25–30% of their labour force in the industrial sector. The low figure for Sudan reflects a lack of industrial development in that country, but this is certainly not the case in Turkey, which has one of the most highly developed and varied industrial sectors in the region.

There is a wide range of values for the proportions employed in service activities. Oil-rich states score high in this respect, as do Israel, Jordan and Lebanon; the very low figure for Sudan is another indicator of that country's low level of development. In most countries there has been a striking increase in service employment over these 20 years, reflecting the development of more sophisticated economies throughout most of the region.

KEY REFERENCES

World Bank, *World Development Report 1984* (World Bank/Oxford University Press, New York, 1984).

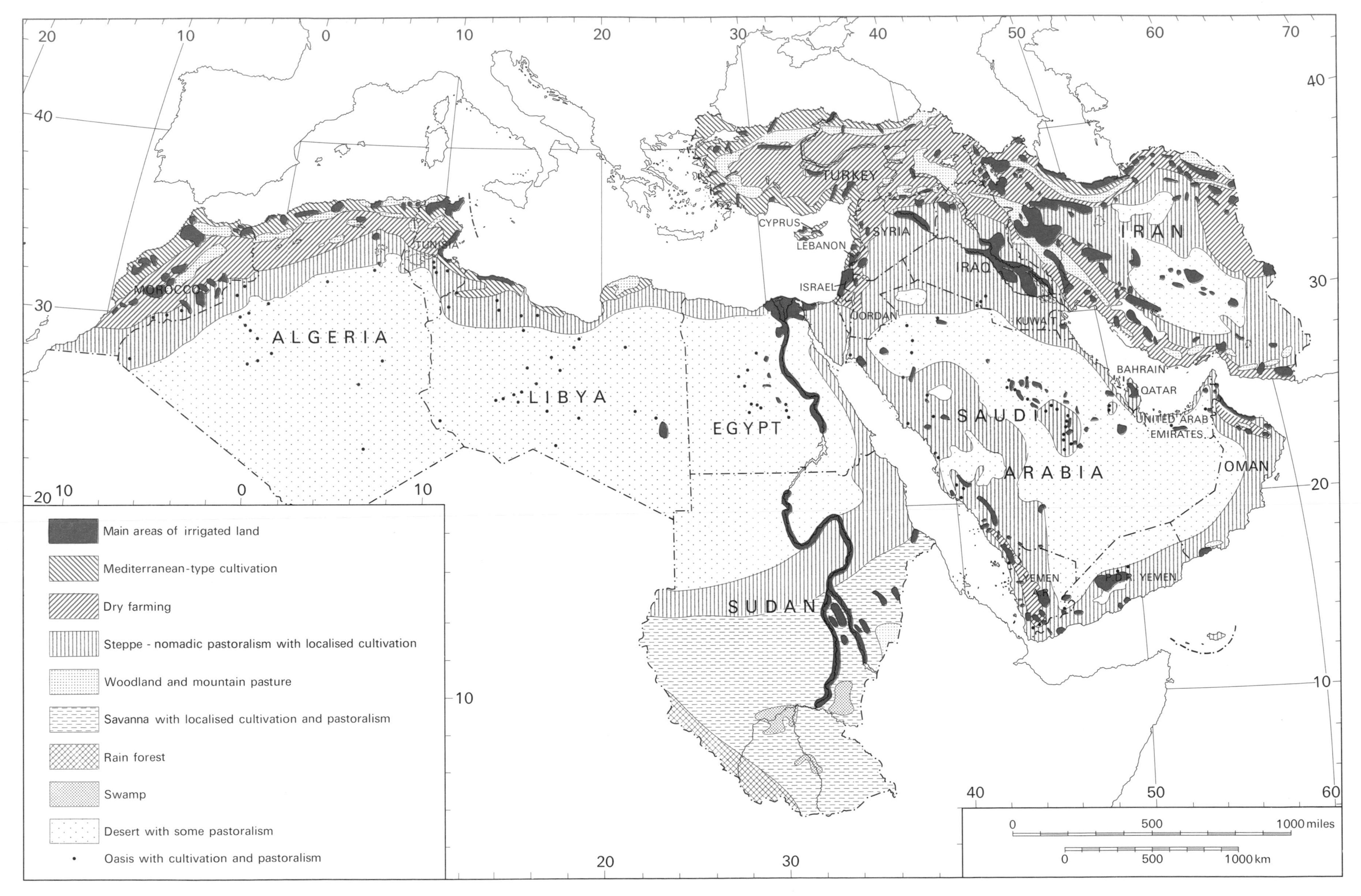

28 Agricultural regions

Agricultural regions

John Dewdney

Despite the industrial and commercial developments discussed elsewhere in this atlas, agriculture remains the predominant economic activity over the Middle East and North Africa as a whole and in the majority of the 22 countries, at least in terms of employment. The economic structure of the population and the contribution of agriculture to the economy of each country are discussed more fully elsewhere (Maps 27 and 39). Suffice it to say that, out of a total of some 80 million, roughly 40% are employed in the agricultural sector. As Table 7 above shows, there are few countries where this proportion is less than 20%; the lowest figures are recorded in oil-rich countries with limited agricultural resources, such as Libya and the Gulf states, or in those few cases where industry and/or commerce play a major role, such as Israel and Lebanon.

The physical environment is the dominant factor in the regional variations in types of agriculture displayed in Map 28. In the first place, rain-fed cultivation is possible only over limited parts of the region, notably throughout most of Turkey, parts of western and northern Iran, along the Levant coast, in high-standing areas of the Arabian peninsula and in the northern Maghreb. Even in these districts, irrigation, while not essential, is often employed as a means of raising crop yields; outside them, it is generally essential. Relief is also a major factor in the northern fold mountain zones. Altogether less than 30% of the land area of the region is utilised for agriculture, some 20% being classed by the F.A.O. as 'permanent meadows and pastures' and only 6.7% as cropland.

Thus there is a primary division on the map between the first three categories, which contain the great bulk of the region's cropland, and the remainder of the region, where cropland may occur in relatively small, scattered patches or not at all, and where the predominant form of land use is pastoralism, which varies in intensity with variations in annual precipitation (see Map 9). As Table 8 shows, this results in a wide variation in the proportion of each country's land which is under crops, a figure which exceeds 30% in only four countries (Tunisia, Turkey, Syria and Cyprus) and is less than 1% in several (Saudi Arabia and the Gulf states).

Three main types of crop farming are distinguished on the map. The most intensive agriculture occurs in areas almost wholly dependent on irrigation. These are of two main types: the riverine plains along the Nile, Tigris and Euphrates in Sudan, Egypt, Iraq and Syria; and the intermontane basins of Iran and the Maghreb. These are not, of course, the only areas in which irrigation takes place; it also occurs widely in areas of predominantly rain-fed cultivation. The largest areas of irrigated land in absolute terms are in fact in Iran, Egypt, Turkey and Sudan, which together have some 70% of all such land in the region; dependence on irrigation is virtually absolute in Egypt and the Gulf states and reaches nearly 50% in Israel.

Table 8. *Agricultural land use*

	Percentage of total land area				
	Cropland				
	Total	Percent irrigated	Pasture	Woodland	Other
North Africa	4.7	14.9	14.9	7.3	73.1
Maghreb	7.0	5.0	17.4	3.4	72.2
Algeria	3.2	4.6	15.3	1.8	79.7
Morocco	18.9	6.3	28.0	11.6	41.5
Tunisia	32.2	3.6	20.2	3.6	44.0
Northeast Africa	3.3	27.0	13.5	9.5	73.7
Egypt	2.5	99.7	—	n	97.5
Libya	1.2	10.9	7.5	0.4	90.9
Sudan	5.2	15.3	23.6	20.3	50.9
Southwest Asia	9.4	16.5	27.4	7.5	55.7
Northern tier	17.0	14.9	22.2	15.9	44.9
Iran	8.4	29.2	26.9	11.0	53.7
Turkey	35.4	7.7	12.2	26.2	26.2
Levant/Mesopotamia	17.0	21.3	17.7	3.2	62.1
Cyprus	46.8	21.8	10.1	18.5	24.6
Iraq	12.5	32.1	9.2	3.5	74.8
Israel	20.7	48.8	40.2	5.7	33.4
Jordan	4.3	9.2	1.0	0.4	94.3
Lebanon	29.1	28.9	1.0	6.9	63.0
Syria	31.5	9.6	45.2	2.7	20.6
*Arabian peninsula**	1.4	18.2	34.1	1.9	62.6
Bahrain	3.2	100.0	6.5	—	90.3
Kuwait	0.1	100.0	—	7.5	92.4
Oman	0.2	92.7	4.7	—	95.1
Qatar	0.3	100.0	4.5	—	95.2
Saudi Arabia	0.5	35.3	39.5	—	60.0
United Arab Emirates	0.2	35.7	2.4	—	97.4
Yemen A.R.	14.3	8.8	35.9	8.2	41.6
Yemen, P.D.R.	0.6	33.8	27.2	7.3	64.9
Total*	6.7	17.3	20.3	7.4	65.6

Note: *Totals exclude Bahrain, Oman, Qatar and U.A.E., for which comparable data were not available.
Dashes indicate data not recorded. n = negligible.
Source: Compiled from data in F.A.O., *Production Yearbook 1983* (U.N.F.A.O., Rome, 1984).

The most productive predominantly rain-fed agriculture is to be found in the areas labelled 'Mediterranean-type cultivation', which occur in coastal regions of North Africa, the Levant and Turkey and in the Caspian coastal belt of Iran. The most distinctive feature of these areas is the production of tree crops, chiefly olives, fruit and vines. Such crops occupy at least 10% of the cultivated land in Algeria, Libya, Syria and Turkey, for example, and over 20% in Israel and 30% in Lebanon and Tunisia.

'Dry-farming' areas are characteristic of the interior of the Maghreb, the Anatolian plateau of Turkey and the better-watered parts of Iran. Cereals are heavily predominant throughout these zones, often accompanied by livestock production. Productivity is closely related to precipitation and is generally lower than in the 'Mediterranean' zones.

Most of the remainder of the Middle East and North Africa is composed of semi-arid and arid steppe and desert zones in which the amount of cropland is very small and the majority is devoted to cereal production. Most of these areas are, at best, low-grade pasture devoted to very extensive stock-rearing; large areas are virtually useless. The savanna lands of the central and southern Sudan have slightly more cultivated land, but are also devoted mainly to livestock.

In the years since World War II, the cultivated areas of several Middle Eastern countries have been greatly increased, mainly as a result of the expansion of dry farming on to former grazing lands: Turkey and Syria, in particular, experienced this kind of development on a large scale in the 1950s. The possibilities for this kind of agricultural expansion are now virtually exhausted. Future increases in agricultural production – very necessary in view of current population growth rates – must depend mainly on raising yields from existing farmland by means of improvements in farming practice assisted by an increase in the area under irrigation.

KEY REFERENCES

Beaumont, P., Blake, G. H. and Wagstaff, J. M., *The Middle East: A Geographical Study* (Wiley, Chichester, 1976).
Bowen-Jones, H. and Dutton, R. W., *Agriculture in the Arabian Peninsula* (Economist Intelligence Unit, London, 1983).
Fisher, W. B., *The Middle East*, 7th edn (Methuen, London, 1978).

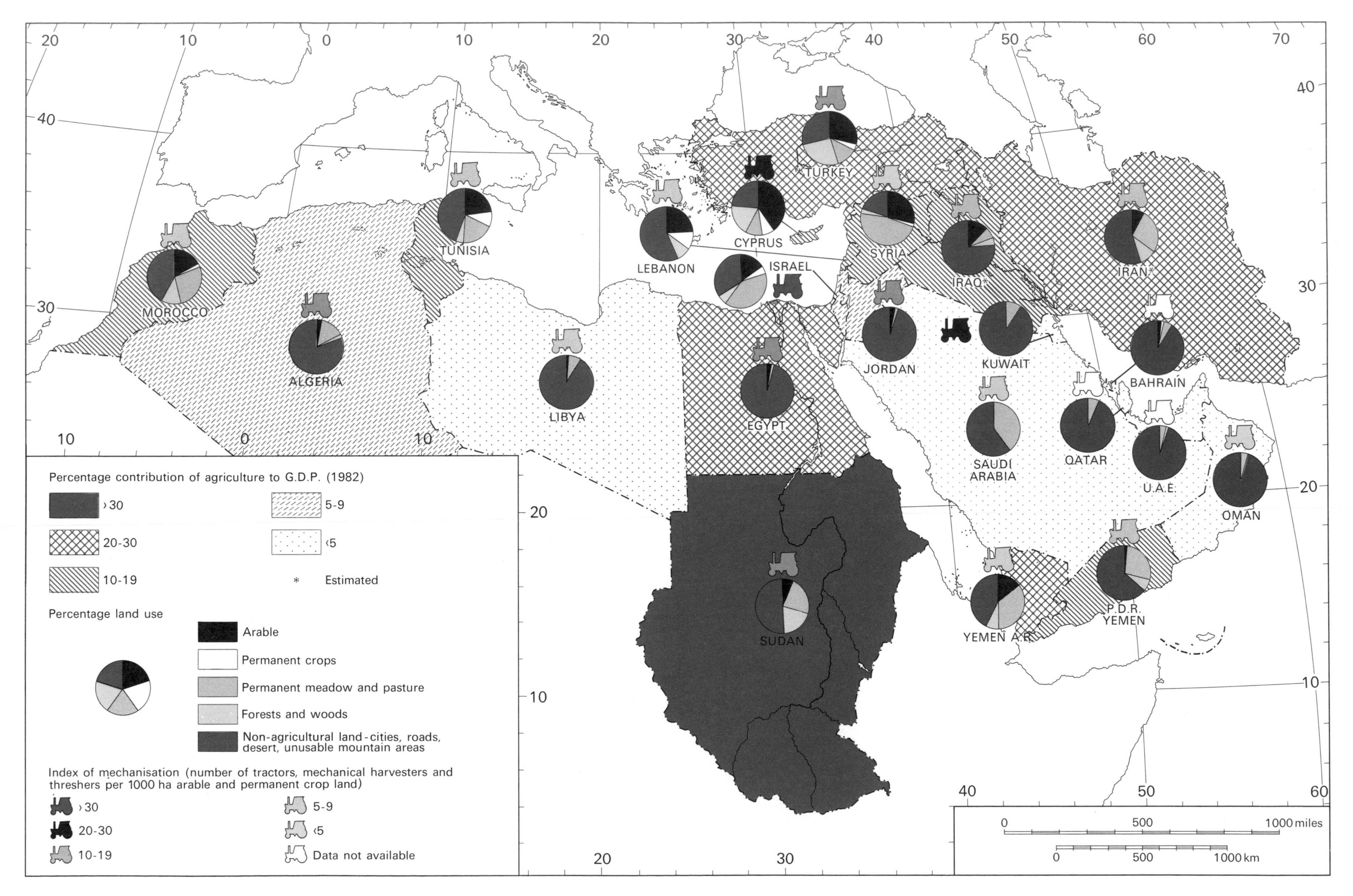

29 Agricultural indicators

Agricultural indicators

John Dewdney

Map 29 covers three important aspects of the agriculture of the Middle East and North Africa, namely (1) the contribution which is made by the agricultural sector to the Gross Domestic Product (G.D.P.) of each country; (2) the use made of the available land resources; and (3) the level of mechanisation achieved, as indicated by the size of each country's stock of tractors and of mechanical harvesters and threshers. None of these measures is wholly satisfactory but here, as in so many other instances, our work has been constrained by the nature of the data available.

The contribution of the agricultural sector to G.D.P.

This is shown by the shading applied to each country. Although, as we have already seen, agriculture employs about 40% of the labour force of the region as a whole and there are few countries in which this proportion is less than 20%, the figures for agriculture's share of G.D.P. are in every case significantly lower than its share of the labour force. In other words, the fraction of the Gross Domestic Product contributed by a worker in the agricultural sector is much smaller than that of a worker employed in a non-agricultural activity such as mineral extraction, manufacturing or the service sector. This, in turn, is reflected in pronounced income differentials between the rural/agricultural population and the predominantly urban population.

The proportion of G.D.P. derived from the agricultural sector is, at least in the context of the Middle East and North Africa, a somewhat negative measure, in the sense that a high figure represents the failure of the country concerned to develop other activities, such as extractive or manufacturing industry or the wide range of service activities associated with economies which have developed beyond the traditional, predominantly agricultural stage. This can be seen from the fact that there is a fairly close negative relationship between the proportion of the G.D.P. derived from the agricultural sector and the level of G.D.P. per capita. This is most clearly seen at the two extremes: Sudan, for example, the one large country in the region where more than 75% of the working population is still employed in agriculture, has, by a considerable margin, the lowest per capita G.D.P., while some of the oil-rich states, where agriculture makes a minimal contribution to G.D.P., have figures for per capita G.D.P. which are among the highest in the world.

Between these two extremes, several countries derive between 20% and 30% of their G.D.P. from the agricultural sector. These include both countries with large resources of agricultural land, exploited for the most part in a somewhat extensive manner, such as Turkey and Iran, and countries with limited land resources worked very intensively, such as Egypt. Agriculture remains important despite the presence of alternative resources – a variety of minerals in Turkey, oil in Iran – which have allowed considerable industrial development. Similarly, in countries where agriculture contributes less than 20% of G.D.P., other activities have increased their contribution in recent decades: oil in Iraq, tourism and other services in Cyprus, manufacturing industry in Israel, for example.

The pattern is a complex one, but in virtually every country both the proportion of the labour force employed in agriculture and that sector's contribution to G.D.P. have declined in recent years. In no way should this be taken as indicating an absolute decline in the value of agricultural production; in practically every country, agricultural output continues to increase. This includes the oil-rich states where part of the wealth derived from oil has been used for agricultural development, including the establishment of intensive irrigated crop production in areas where agriculture was previously confined to extensive livestock rearing.

Land use

The varied uses of the land resources of each country are indicated by pie diagrams. The great range in the size of the countries has, unfortunately, precluded us from making these symbols proportionate to the size of the land area; thus the map shows only the proportions, not the actual areas, devoted to each use. Some data relevant to this aspect have already been provided in Table 8 above. Aspects of the physical environment, especially the water balance as indicated in Map 10, and relief (Map 4), are highly significant.

The most striking feature of the Middle East and North Africa as a whole, and particularly the more southerly countries (excluding Sudan), is the large proportion of the total land area classed as 'non-agricultural'; arable land, land under perennial crops and permanent meadow and pasture together constitute barely 30% of the land in the region. Only in a handful of countries – Tunisia, Turkey, Cyprus, Israel, Syria and Yemen A.R. – is non-agricultural land less than half the total, and the proportion not used rises to 80% or 90% in territories which are predominantly desert, such as Libya, Egypt and the Gulf states. In most of these cases (Egypt is the outstanding exception) the bulk of the land which is used is utilised only for extensive livestock grazing.

Within the areas devoted to agriculture, the amount which is actually used for crops, both annual and perennial, also varies greatly from country to country. Cropland of both types together exceeds 30% of the total land area in only five cases: Cyprus, where the figure is 47%, Turkey, Syria, Lebanon and Tunisia; proportions of 10–20% occur in four more: Morocco, Iraq, Israel and Yemen A.R. Thus 13 of our 22 countries have less than 10% of their land area under crops, the proportion falling below 1% in Kuwait, Oman, Qatar, Saudi Arabia, the United Arab Emirates and P.D.R. Yemen.

In the great majority of countries – Tunisia, Egypt, Turkey, Cyprus, Iraq, Lebanon and Jordan are the exceptions – grazing land is considerably more extensive than cropland. Care is needed here, however, since the data on which the map is based are dependent on the definitions of land use categories adopted by the various countries. Saudi Arabia, for example, reports nearly 40% of its total land area as being occupied by permanent pasture whereas Egypt gives no figure for this category, yet Egypt contains considerable areas of grazing land which are no worse than those of much of the Arabian peninsula.

Similar reservations must apply to the areas returned as 'forest and woodland', which in any case exceed 10% of the land area only in Sudan, Turkey, Morocco, Iran and Cyprus; in several cases special laws regarding cutting and grazing apply to land officially designated 'forest' and the area so designated may be increased by law or decree. In the case of Turkey, for example, a quarter of the territory is, officially, forest and the area in this category was doubled during the 1960s and 1970s. Most of the new 'forest' land, however, is degenerate scrub and only about a quarter of the land in this category can be classed as true forest.

Mechanisation

The level of mechanisation is indicated only in a very general way by the index used here, which is the number of tractors and mechanical harvesters and threshers per 1000 ha of arable and permanent crop land. One problem, for example, is the relatively high level of mechanisation in Israel, which recorded 63 mechanisation units per 1000 ha in 1983, approximately double that of any other state. There are also several countries where the appropriate data are not available. For the region as a whole, the average value is low – about ten units per 1000 ha – when compared with levels of mechanisation in the developed world (e.g. U.S.A. 28, U.K. 85). This average is significantly exceeded in only half a dozen countries: Egypt, Turkey, Cyprus, Jordan, Algeria and, above all, Israel, where the figure is more than six times the regional average.

KEY REFERENCES

F.A.O., *Production Yearbook 1983* (U.N.F.A.O., Rome, 1984).

World Bank, *World Development Report 1984* (World Bank/Oxford University Press, New York, 1984).

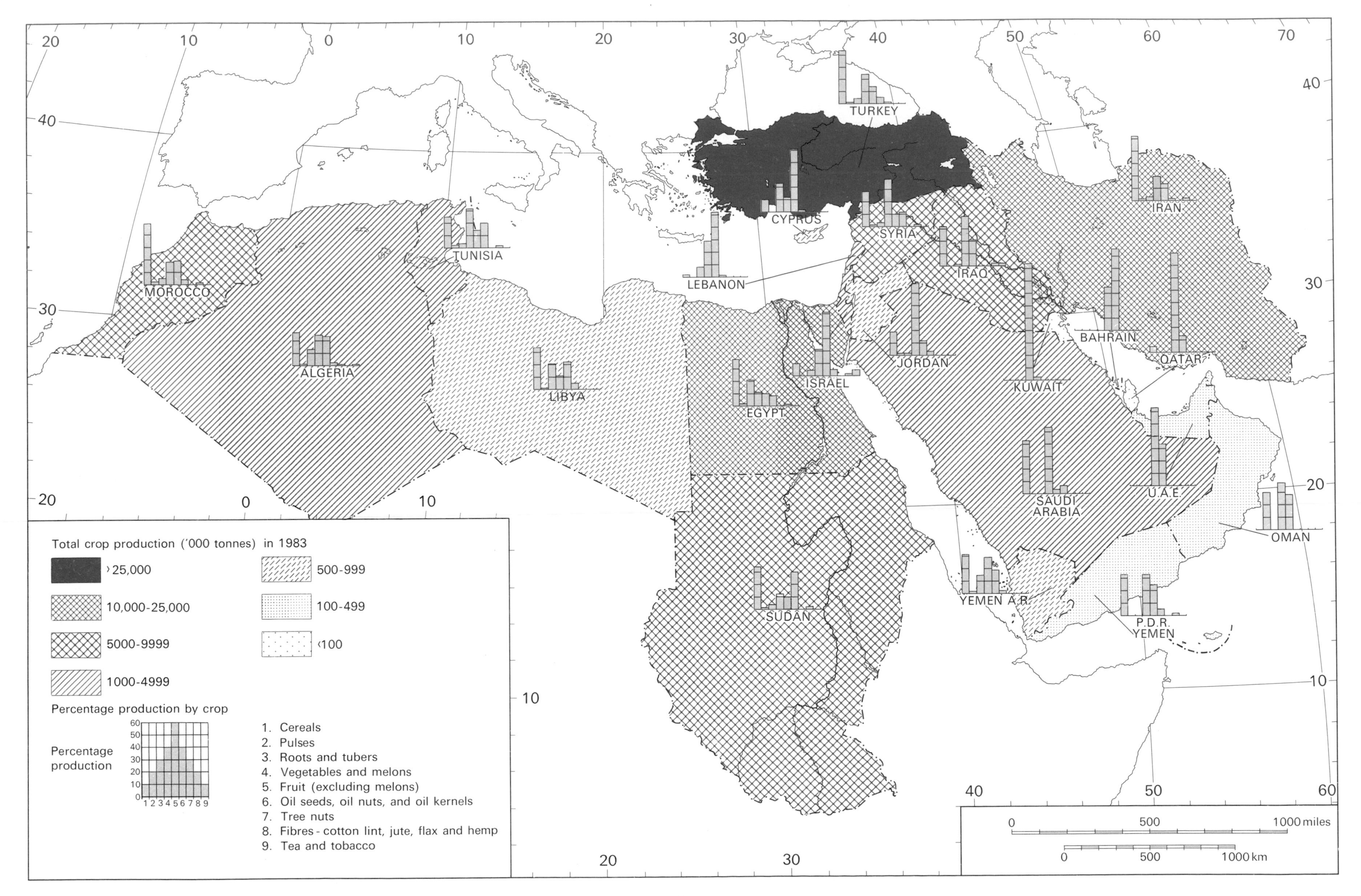
TURKEY
IRAN
CYPRUS
SYRIA
TUNISIA
IRAQ
LEBANON
MOROCCO
BAHRAIN
QATAR
JORDAN
ALGERIA
ISRAEL
LIBYA
KUWAIT
EGYPT
SAUDI ARABIA
U.A.E.
OMAN
YEMEN A.R.
SUDAN
P.D.R. YEMEN
Total crop production ('000 tonnes) in 1983
› 25,000
10,000-25,000
5000-9999
1000-4999
500-999
100-499
‹100
Percentage production by crop
Percentage production
1. Cereals
2. Pulses
3. Roots and tubers
4. Vegetables and melons
5. Fruit (excluding melons)
6. Oil seeds, oil nuts, and oil kernels
7. Tree nuts
8. Fibres - cotton lint, jute, flax and hemp
9. Tea and tobacco
0 500 1000 miles
0 500 1000 km

30 Crops

Crops

John Dewdney

Map 30 shows both the total volume of crop production and the composition of that production in terms of the nine groups of crops indicated in the key. Of a total output well in excess of 120 million tonnes a year, about one-third comes from the North African section and two-thirds from the Asian countries of the Middle East. Among the latter, the northern tier countries – Turkey and Iran – are overwhelmingly predominant: together they are responsible for more than 40% of the output of the entire region, Turkey alone contributing nearly 30%. At the other extreme, all the countries of the Arabian peninsula combined account for only about 5% of the total crop production.

As the map indicates, the composition of this production varies considerably from country to country, reflecting the regional variations in the nature of agriculture which have already been discussed in relation to Map 28. In areas where extensive rain-fed cultivation is possible cereals are predominant to a very pronounced degree, accounting for more than 40% of total crop production in Morocco, Turkey and Iran. The latter two, although they cover only 16.9% of the land area and contain only about one-third of the total population of the Middle East and North Africa, are together responsible for nearly 60% of the region's grain production, 43% of which comes from Turkey. These countries, at least, are in the fortunate position of having an assured and adequate supply of the most basic of foodstuffs: bread. Indeed, there have been times in her recent history when Turkey, virtually alone among the countries of the region, has achieved a sizeable net export of food grains, though her rapid population growth has prevented grain exports from becoming a permanent feature of her economy. Several other countries show between one-fifth and one-third of their crop production in the form of grain. These include some – like Syria and Algeria – with important areas of rain-fed farming, and others – like Egypt and, to a lesser extent, Iraq – which have, perforce, devoted much of their fertile irrigated land to cereal production. Others, such as Saudi Arabia and the Yemens, also have at least 30% of their output in the form of cereals, but actual quantities are very small. There are other cases – Kuwait, Bahrain and the United Arab Emirates, for example – where the physical environment precludes any possibility of feeding the population from the minuscule land resources at their disposal, which are therefore devoted almost entirely to the production of fruit and vegetables, usually under irrigation.

Of the 60 million tonnes of grain produced in the region each year, at least three-quarters is in the form of the two main food cereals, wheat and barley. At the country level, these constitute more than 85% of total cereal output in the great majority of cases and over 90% in several. The main exceptions are Egypt, where maize is the dominant cereal crop over large areas, and Sudan and Yemen A.R., where sorghum and millet predominate. It is important to remember that we are discussing percentage rather than absolute figures: in Turkey, for example, maize represents only 6% of total cereal production, but the amount actually produced is second only to Egypt.

With a few minor exceptions, then, cereals are predominant in crop production at the national level and the most common cropping cycle is a three-year one, comprising two years of wheat or barley and one year fallow. At the sub-national level, however, there are few regions or districts in which cereals are the only crops, though it is also rare for them not to occupy at least half of the cultivated area.

Pulses – a variety of leguminous crops used primarily in feeding livestock – make up only a very small proportion of total crop production and are associated mainly with dry-farmed cereal cultivation. Their relatively small contribution reflects a general lack of integration between the crop and livestock sides of traditional agriculture in the region; animals have generally been fed on fallow land and permanent pasture rather than on crops grown specially for the purpose. Attempts by governments and international agencies to increase fodder crop production as a means of raising the output of livestock products have met with only limited success.

Root and tuber crops with their heavy water demands and relatively low value/weight ratio are also of minor importance in all but a few countries. Cassava and yams are produced in southern Sudan, but the most widely grown of these crops is the potato. A few countries – the Maghreb states, Egypt and Cyprus – sell their 'early' potatoes on the European market.

Vegetable crops constitute 20–30% of agricultural production in the great majority of countries. They have long been grown on a small scale in gardens in and around settlements, but their production as field crops has increased considerably in recent decades. Whereas in the past vegetables were produced mainly for local domestic consumption, urban growth has produced a rising demand and improvements in communications have permitted the development of specialised, commercial vegetable production. They are most important in the Maghreb states, Israel, Lebanon and Jordan, all of which have developed export markets. As a proportion of total agricultural output, vegetables are dominant throughout the Gulf states where satisfying the local demand for fresh produce is the most that can be achieved.

A rather similar pattern can be seen in the case of fruit production. Fruit of various kinds is produced on a considerable scale in the great majority of countries, the actual type being strongly influenced by environmental and economic factors. Temperate fruits are particularly associated with upland areas, citrus fruits with coastal lowlands. Vines are found mainly in the Maghreb and Turkey. In at least three countries – Cyprus, Lebanon and Israel – fruits contribute some 50% of agricultural production and are exported on a considerable scale.

The category 'oil seeds, oil nuts and oil kernels' includes a variety of items ranging from sunflower and sesame in the temperate zone to ground nuts in the Sudan, but the most important item is the olive, grown mainly in the Mediterranean section of the region. The main 'tree nut' is the date, a traditional crop of oases in the arid regions, now produced commercially in several countries. Among the fibre crops, by far the most important is the cotton of Egypt, Sudan, Iran, Syria and Turkey. Tea and tobacco are produced mainly in Turkey and Iran, though several other countries now have small areas of tobacco.

The crops in these last three categories (7–9) are produced in relatively small quantities when compared with the enormous weight of cereal production and thus do not show up well on the bar graphs, which are based on the volume of production. There is a clear distinction between these crops and such bulky items as cereals, roots and tubers and, indeed, vegetables. If data were available, diagrams based on the value rather than the weight of production of the various commodities would give some interesting results.

KEY REFERENCES

Beaumont, P. and McLachlan, K. S. (eds.), *Agricultural Development in the Middle East* (Wiley, Chichester, 1985).

Economic Intelligence Unit, *Oxford Regional Economic Atlas of the Middle East and North Africa* (Oxford University Press, London, 1960).

F.A.O., *Production Yearbook 1983* (U.N.F.A.O., Rome, 1984).

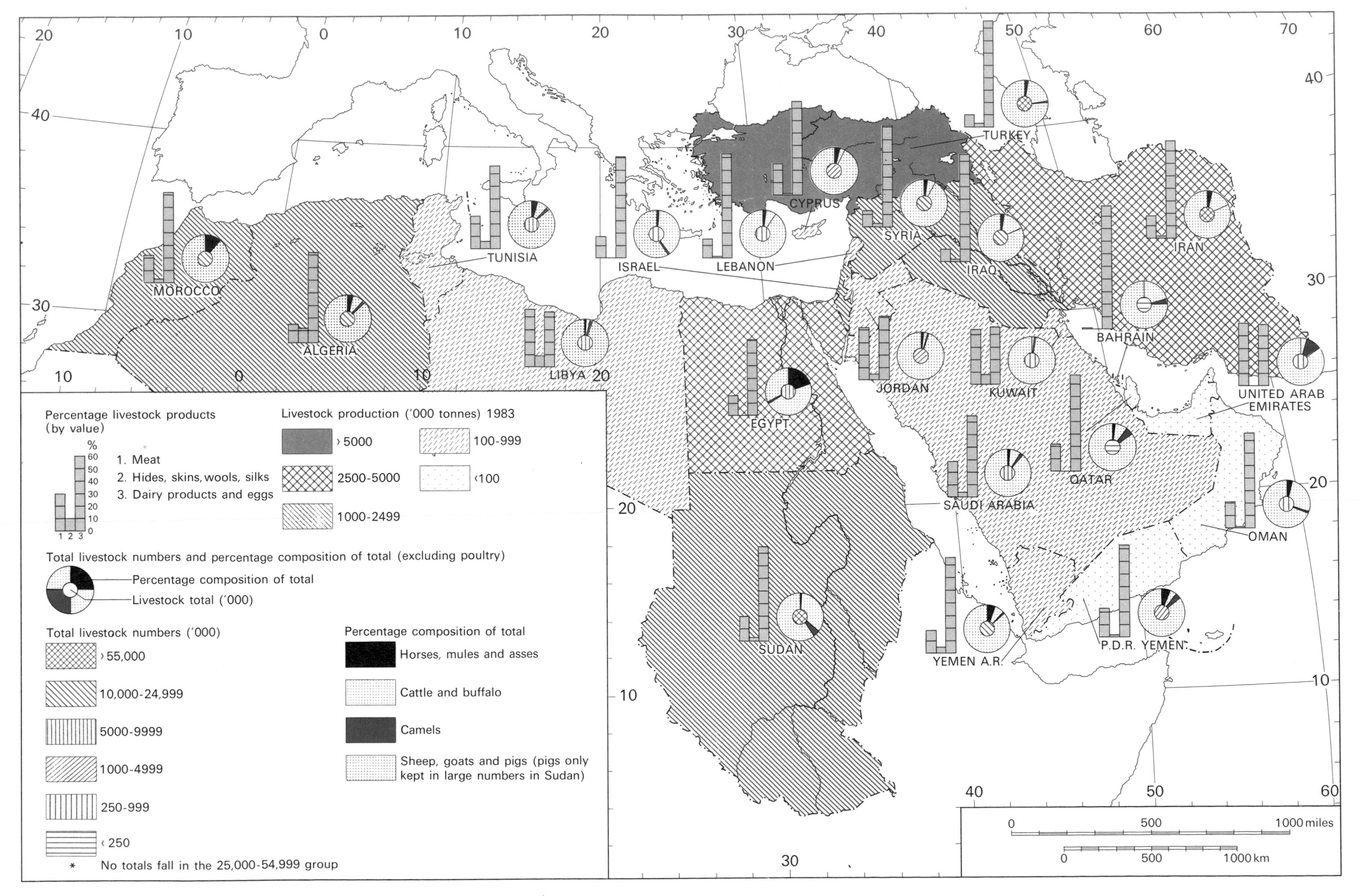

31 Livestock

Livestock

John Dewdney

Livestock rearing has always been a major element in traditional agricultural economies throughout the Middle East and North Africa. However, in contrast to farming systems in northwest Europe and other temperate lands, where crop and livestock production are closely integrated and both crop and livestock products contribute to the income of the majority of farms, mixed farming is rare in this region. Crop and livestock farming are often spatially separated and are commonly carried out by separate groups of cultivators and pastoralists. This is particularly the case where, as in the semi-deserts of North Africa and the Arabian peninsula, livestock rearing is virtually the only activity and an element of true nomadism survives, but it also applies to environmentally more favoured areas where both types of agriculture are carried on side by side but separately. In the latter case, areas of strong relief offer alternative grazing lands in the summer and winter seasons and transhumance is a common feature. There is a high demand for meat in Islamic communities, particularly at festival times, and very large quantities of mutton and beef are imported, notably to Saudi Arabia.

As Table 9 shows, of an estimated 330 million livestock kept within the Middle East and North Africa as a whole, about three-quarters are sheep, goats and pigs; pigs are found in significant numbers only where Jewish and Islamic dietary constraints do not apply, as in Cyprus, parts of Lebanon and above all in Christian areas of Sudan, and number less than half a million. A further 20% of all livestock are cattle and buffaloes, the latter significant only in Egypt, Iran and Turkey; about 4% horses, mules and asses, and only 1% camels. Thus cattle, sheep and goats together constitute rather more than 90% of all livestock.

In terms of absolute numbers, sheep and goats (*c.* 250 million) outnumber cattle (*c.* 60 million) by about four to one. These two groups of animals, however, make very different demands on land resources in terms of their grazing and fodder crop requirements and yield very different quantities of livestock products. Thus it is common practice to convert absolute numbers of livestock into 'livestock units', making one head of cattle equivalent to six sheep or goats. When this is done, and ignoring the other two categories of livestock, which are essentially draught animals, cattle, sheep and goats total about 100 million units, of which cattle constitute some 60%.

The circular symbols on the map (which are based on absolute numbers) show that there is considerable variation in the composition of the livestock population. One feature common to the great majority of countries is the heavy predominance of sheep and goats which in 18 out of 22 cases constitute more than 75% of all livestock, the figure rising above 90% in six countries – Libya, Cyprus, Jordan, Lebanon, Syria and Kuwait. The proportion of cattle is lowest in countries with large areas of semi-arid grazing land suitable only for the rearing of sheep and goats, as in the Maghreb states and the Arabian peninsula.

A few countries have exceptionally high proportions of cattle. Most striking in this respect is Egypt, where cattle and buffalo constitute 47% of all livestock – and Egypt has two-thirds of all the buffalo in the region. Because of the particularly intense pressure of population on limited land resources in Egypt, crop production dominates the agricultural economy. In recent decades, the crop rotations used have increasingly involved nitrogen-fixing fodders, such as alfalfa, which permit the support of large numbers of cattle and water buffalo. A high proportion of these are used as draught animals and yields of meat and dairy products are comparatively low, a situation which prevails over much of the region but is particularly marked in this case. Sudan, in this respect as in so many others, is in a class of its own. Cattle (about one-third of all livestock) are kept by most farming families as a source of manure, labour, food and, in many areas, prestige. Finally, Israel, where cattle are 42% of all livestock, is virtually alone among the countries of the region in having developed mixed farming systems which involve the widespread use of fodder crops in addition to very limited areas of natural grazing.

Table 9. *Livestock numbers (000)*

	Horses, mules and asses	Cattle and buffalo	Camels	Sheep, goats and pigs	Total
Algeria	937	1,400	154	16,540	19,031
Morocco	2,321	3,000	240	21,281	26,842
Tunisia	334	560	175	6,024	7,093
Egypt	1,785	4,219	80	2,907	8,991
Libya	74	200	135	6,300	6,709
Sudan	709	19,550	2,500	32,400	55,159
Iran	2,273	8,820	27	48,330	59,450
Turkey	2,365	17,908	10	67,863	88,146
Cyprus	49	43	0	1,060	1,152
Iraq	543	3,240	250	15,800	19,833
Israel	11	330	11	450	802
Jordan	28	40	15	1,500	1,583
Lebanon	16	50	0	599	665
Syria	336	802	7	12,100	13,245
Bahrain	—	6	1	22	29
Kuwait	—	18	5	850	873
Oman	23	150	6	390	569
Qatar	2	10	6	110	128
Saudi Arabia	119	500	160	5,800	6,579
U.A.E.	—	30	70	540	640
Yemen A.R.	743	950	108	10,650	12,451
Yemen, P.D.R.	170	120	100	2,350	2,740
Total	12,838*	61,946	4,060	253,866	332,710

Note: *Total excludes Bahrain, Kuwait and U.A.E., for which comparable data were not available. Dashes indicate data not recorded.
Source: F.A.O., *Production Yearbook 1983* (U.N.F.A.O., Rome, 1984).

The actual numbers of livestock kept in each country are, of course, closely related to the land resources available to support them. Thus about 26% of all the livestock in the region is to be found in Turkey, about 18% in Iran and 17% in Sudan. The nearest rivals are Morocco, Algeria and Iraq, with 6% to 8% in each. In terms of livestock units per capita, Sudan stands head and shoulders above all other countries of the region, and Turkey is also well above average. Very low values are recorded in Egypt, Libya, Israel, Jordan and Lebanon and all the countries of the Arabian peninsula except Yemen A.R. In recent decades the number of camels and goats has fallen sharply in some states with the decline of nomadism.

The map also shows, by a system of shading, the total output of livestock products of each country. Not surprisingly, this is related both to territorial size and to land resources. Thus Turkey, with its relatively favourable climate, large land resources and lack of extensive negative areas takes the lead with a production well in excess of five million tonnes. Most of the other large countries of the region – Morocco, Algeria, Egypt, Sudan and Iran – together with Syria and Iraq produce between one and five million tonnes. Relatively intensive production from a small land area places Israel in the same category. Libya and Saudi Arabia, on the other hand, with their large areas of desert and small livestock populations produce less than a million tonnes each and fall into the same class as much smaller countries like Kuwait, Lebanon and Cyprus. Bahrain, the United Arab Emirates, Oman and P.D.R. Yemen each produce less than 100,000 tonnes.

KEY REFERENCES

Beaumont, P. and McLachlan, K. S. (eds.), *Agricultural Development in the Middle East* (Wiley, Chichester, 1985).
Clawson, M., Landsberg, H. H. and Alexander, L. T., *The Agricultural Potential of the Middle East* (Elsevier, New York, 1981).
F.A.O., *Production Yearbook 1983* (U.N.F.A.O., Rome, 1984).

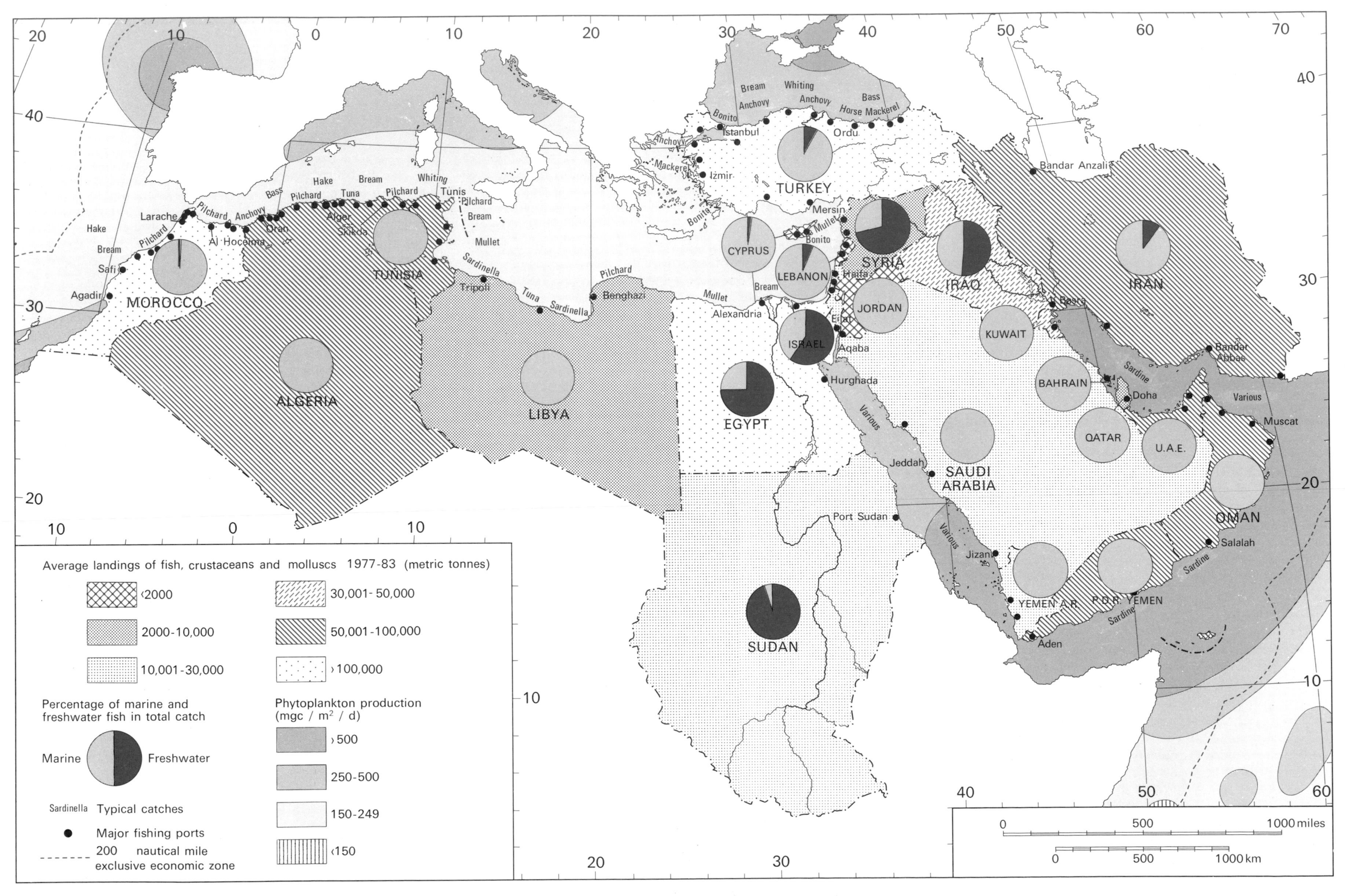
MOROCCO
ALGERIA
TUNISIA
LIBYA
EGYPT
SUDAN
TURKEY
CYPRUS
LEBANON
SYRIA
ISRAEL
JORDAN
IRAQ
IRAN
KUWAIT
BAHRAIN
QATAR
U.A.E.
SAUDI ARABIA
OMAN
YEMEN A.R.
P.D.R. YEMEN
Larache
Safi
Agadir
Al Hoceima
Oran
Alger
Skikda
Tunis
Tripoli
Benghazi
Alexandria
Istanbul
Izmir
Ordu
Mersin
Haifa
Eilat
Aqaba
Hurghada
Jeddah
Port Sudan
Jizan
Aden
Salalah
Muscat
Doha
Basra
Bandar Anzali
Bandar Abbas
Hake
Bream
Pilchard
Anchovy
Bass
Tuna
Whiting
Mullet
Sardinella
Bonito
Mackerel
Horse Mackerel
Sardine
Various
Average landings of fish, crustaceans and molluscs 1977-83 (metric tonnes)
‹2000
2000-10,000
10,001-30,000
30,001-50,000
50,001-100,000
›100,000
Percentage of marine and freshwater fish in total catch
Marine
Freshwater
Phytoplankton production (mgc / m² / d)
›500
250-500
150-249
‹150
Sardinella Typical catches
Major fishing ports
200 nautical mile exclusive economic zone
0 500 1000 miles
0 500 1000 km

32 Fishing

Fishing

Gerald Blake

The most important form of plant life in the oceans is the phytoplankton, microscopic algae which provide the lowest level of the food chain which leads to the support of fish. The abundance of phytoplankton depends on many factors, including water temperature and the supply of nutrients in the surface water where light is sufficient for photosynthesis. Such nutrients may be brought to the surface by storms, currents or, locally, by river action. Phytoplankton production in the seas adjacent to the Middle East and North Africa is generally modest, particularly in the Mediterranean where it is predominantly less than 250 milligrams of carbon per square metre per day (mgc/m^2/d). The region's most productive waters are off the coasts of southern Arabia and northwest Africa (just shown on the map off Western Sahara). Here, upwelling due to the prevailing wind systems brings nutrients to the surface and phytoplankton production exceeds 500 mgc/m^2/d. No data was available for the Caspian Sea, which provides Iran with a sizeable harvest of sturgeon from which caviar is obtained.

By world standards, marine fish catches in the region are unremarkable, amounting to less than 2% of the world total from rather less than 2% of the oceans by area. Turkey (438,000 tonnes in 1981) and Morocco (381,000 tonnes in 1981) have by far the largest catches by volume in the region, but rank 15th and 19th respectively in the world. About 95% of Turkey's catch is from the Black Sea, and 90% of Morocco's from the Atlantic Ocean. The Mediterranean is often perceived by visitors to be a rich fishing area, but catches are relatively small although the variety of edible species marketed is very great. Little mixing of Mediterranean waters takes place to bring nutrients to the surface and there are few rivers contributing nutrients from the land, especially along the coast of North Africa. Completion of the Aswan Dam in 1970 deprived the eastern Mediterranean of valuable nutrients which formerly supported sardine fisheries. Pollution has also taken its toll in reducing fish stocks in places. Nevertheless, most states take a keen interest in fishing, and Mediterranean fish fetch among the highest prices in the world. Fish are a very important element in the cultural life of the Mediterranean world. Thus big efforts are being made to improve fishing techniques and marketing methods, and Mediterranean Sea catches have steadily increased to over one million tonnes, or about 1.5% of the world catch. Certain species are already overfished, however, and national fishing grounds are being more jealously guarded.

Fishing is unimportant in the Red Sea, although nutrient-rich waters from the Arabian Sea penetrate the southern Red Sea. The total Red Sea catch is about 70,000 tonnes, equivalent to about one-sixth of Turkey's annual catch, and prospects for increasing landings are not good. Several attempts to introduce commercial fishing in the Red Sea have failed. Oman and P.D.R. Yemen on the other hand both land about 70,000 tonnes per annum, exploiting the relatively abundant fish stocks of the Arabian Sea. In the past both states have granted concessions for foreign fishing vessels from Korea, Iraq, Japan and the Soviet Union to fish within their 200 nautical mile exclusive economic zones. The volume of fish caught in the Gulf (over 100,000 tonnes) is surprisingly small. The Gulf has a long tradition of artisanal fishing and pearling. Pearl fishing centred on Bahrain was once a major source of income, but died out between the world wars because of competition with Japanese cultured pearls. In the 1950s and 1960s, commercial shrimp fishing grew rapidly, peaked, and then declined, probably as a result of over-fishing. Gulf shrimps were exported to Japan, Europe and the United States from Kuwait and Bahrain, but are now largely confined to Middle Eastern markets. Gulf fisheries were severely damaged in the short term by the disastrous oil slick resulting from Iran–Iraq hostilities in 1983.

Catches are achieved in three contrasting types of fishing. First, freshwater fishing is significant in Sudan, Egypt, Syria, Iraq and Israel where fish from rivers, lakes and fish farms make up more than half the national catch. Rural populations often benefit most from inland freshwater catches. Intensive aquaculture (fish farming) is highly developed in Israel, and might be increased in certain other states of the region. Secondly, the most widespread type is traditional or artisanal fishing and its derivatives. It includes a bewildering variety of boats and fishing gear, but the majority of boats are under 100 gross tonnage, only suitable for fishing relatively close to the shore. Typically, traditional fishing is seasonal and catches are marketed locally. Boats are becoming increasingly motorised, often with the help of government subsidies. Several thousand boats are engaged in this type of fishing. Thirdly, there are a growing number of ocean-going, modern fishing vessels of over 500 gross tonnage designed to undertake large-scale commercial fishing, often involving trawling, and voyages of hundreds of kilometres from port. Many states are investing heavily in commercial fishing enterprises, though not always with great success. There are about 400 fishing vessels of over 500 gross tonnage belonging to states of the Middle East and North Africa, the largest fleets being those of Kuwait and Morocco. These larger vessels with greater range than inshore boats are most likely to become engaged in disputes over access to fishing grounds. Finally, recreational fishing from small boats and from the shore is a common pastime, particularly in the Mediterranean, but the size of catch goes unrecorded.

Anxiety over the protection of fish stocks has been one of the reasons for the extension of territorial water claims and the introduction of exclusive economic/fishing zones in recent years. The 1982 United Nations Convention on the Law of the Sea gives coastal states exclusive control of all resources up to a distance of 200 nautical miles offshore. Only P.D.R. Yemen, Oman and Morocco have their full entitlement of exclusive economic zone because of the narrowness of the region's seas. In the Mediterranean sea the introduction of E.E.Z.s will exclude fishermen of Greece, Italy and Spain from traditional fishing grounds off North Africa. Japanese tuna boats which fish in the Mediterranean will also be excluded. A number of serious fishing disputes have already occurred in the region, notably between Spain and Morocco and Italy and Tunisia.

Fishing ports are indicated on the map, but are difficult to define and many more might be shown. Few ports are exclusively devoted to fishing, apart from extremely small bays and harbours, too numerous to show on the map. Large fishing vessels frequently operate out of commercial ports where they may represent a relatively minor activity. Such ports provide the best facilities for landing and marketing large quantities of fish. In the region as a whole, the storing, processing, and marketing of fish are not well developed and several states include investment in these activities in their current development plans. The average fish consumption per capita is still low, and in certain inland regions it is negligible. It is doubtful whether local supplies will be able to meet future increases in demand.

KEY REFERENCES

Borgese, E. M. and Ginsberg, N. (eds.), *Ocean Yearbook 4* (University of Chicago Press, Chicago, 1983).

Couper, A.D. (ed.), *The Times Atlas of the Oceans* (Times Books, London, 1983).

General Fisheries Council for the Mediterranean, *Statistical Bulletin No. 5* (F.A.O., Rome, 1984).

Other statistics from unpublished F.A.O. database, 1985.

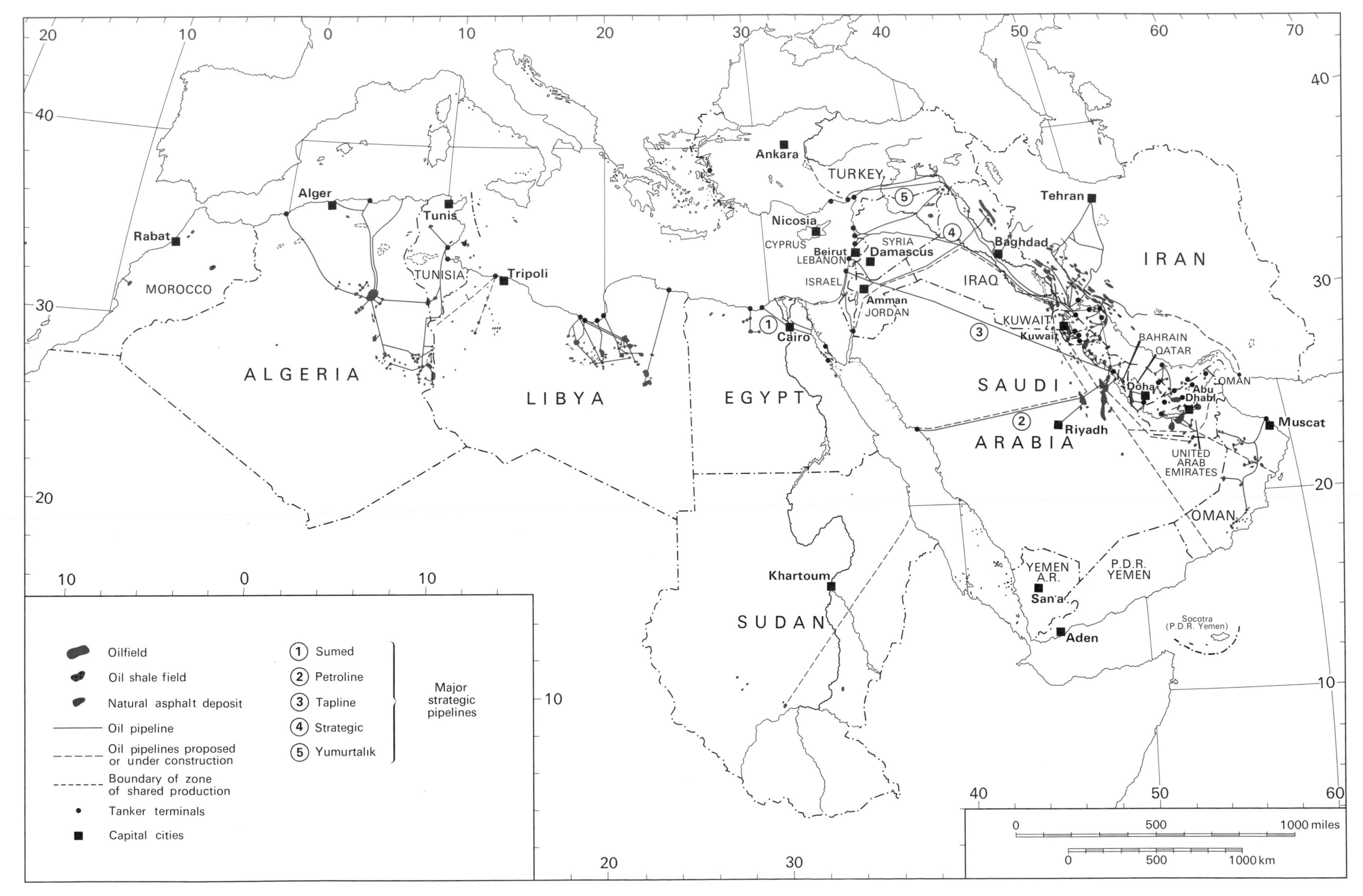

33 Oilfields and pipelines

Oilfields and pipelines

Alasdair Drysdale

The Middle East and North Africa have 61.4% of the world's proven oil reserves as well as the world's highest reserve-to-production ratio (see Table 10). In addition, 9 of the world's 10 largest oilfields and 28 of the 33 super-giant fields that have over half of the world's oil are located in the region. Saudi Arabia's Ghawar field, the world's largest, alone has twice as much oil as the entire proven reserves of the United States. Most of these vast fields are concentrated within the so-called Arabian–Iranian oil province, an area measuring some 1200 by 800 kilometres around the head of the Persian/Arabian Gulf. Consequently, Saudi Arabia, Kuwait, Iran, and Iraq between them have roughly 80% of the region's reserves, while many countries have little or no oil.

Because so much of the Middle East's oil lies within a few giant shallow fields within a small area and close to marine terminals on the Gulf, exploration and production costs have been exceptionally low. Moreover, geology and undivided ownership permit extraordinarily high production rates for individual wells: the typical Iraqi well yields 13,480 barrels of oil daily (b/d), for example, its American counterpart only 17.

As recently as the early 1940s, the Middle East and North Africa produced only 5% of the world's oil. By 1979, they produced 40%. However, total global demand has dropped sharply since then because of economic recession and, more particularly, a sharp increase in oil prices in 1979. Simultaneously, exploration and production outside the Middle East have risen dramatically. As a result, Middle East and North African production fell from 25.1 million b/d in 1979 to only 14.8 million b/d in 1984 – about one-quarter of the world total. The biggest fluctuations in production have been in Saudi Arabia, which acts as a swing producer able to exercise some production control by raising or lowering production within O.P.E.C. In 1980, Saudi production approached 10 million b/d, almost half of the region's total output and roughly four times more than Iraq, the second largest producer. By 1984, Saudi production had fallen to only 4.7 million b/d, still the region's highest. Production has also fluctuated widely in Iran, plummeting from 5.2 million b/d in 1978 to only 1.3 million b/d in 1981 as a result of the 1979 revolution. It had climbed back to 2.2 million b/d by 1984, making it the second largest producer. Elsewhere in the Middle East and North Africa swings in production have generally been less extreme. Algeria, Egypt, Iraq, Kuwait, Libya and the United Arab Emirates each produced between 900,000 b/d and 1.2 million b/d in 1984. It must be emphasised that the impact of oil has been greatest in countries producing most oil on a per capita basis, not in those producing most oil. Thus, 1983 per capita oil exports amounted to $451 in Iran, $519 in Algeria, $4,385 in Saudi Arabia, $8,387 in the United Arab Emirates, and $10,103 in Qatar.

Although the Middle East and North Africa are major producers of oil, they are not major consumers. Most of their oil is shipped to Western Europe, Japan and North America by tanker. Accordingly, an elaborate network of pipelines links the oilfields with marine export terminals, loading platforms and refineries. Of particular interest are the strategic pipelines that greatly shorten tanker voyages and, in certain cases, by-pass shipping routes which are perceived to be vulnerable to political interference. In the past, the most important of these was the 1718-km Tapline, which transported up to 500,000 b/d from eastern Saudi Arabia via Jordan, the Golan Heights, and Syria to Sidon in Lebanon and cut 3200 km off the tanker trip to Europe (far more after the closure of the Suez Canal in 1967). However, Tapline has been vulnerable to closure because it passes through a major war zone and has been the subject of frequent disputes over transit fees. No oil has been exported from the region through Tapline for several years. Of far greater significance is Saudi Arabia's 1213-km Petroline, which opened in 1981 and connects the eastern oil fields with Yanbu on the Red Sea. Present capacity of 1.85 million b/d may eventually be doubled. When used in conjunction with Egypt's 335-km 1.6 million b/d Sumed pipeline, which opened in 1977 and runs between the Gulf of Suez and the Mediterranean, Petroline will allow a substantial proportion of Saudi oil to be shipped out of the Middle East without having to go through the vulnerable Hormuz and Bab al Mandeb straits, the Suez Canal, Syria and Lebanon, or around the Cape of Good Hope. Saudi Arabia, Iraq, the United Arab Emirates and Oman are also exploring the possibility of building a second line that would by-pass the Strait of Hormuz, this one from Ras Tannurah to the Gulf of Oman.

Table 10. *Oil reserves and production*

	Proven reserves (m. barrels)	Share of total (%)	Production (000 b/d)	Share of total (%)
Middle East and North Africa	433,200	61.4	14,875	25.7
Algeria	9,000	1.3	990	1.5
Abu Dhabi	30,500	4.3	840	1.4
Dubai	1,400	0.2	365	0.6
Egypt	3,200	0.5	915	1.6
Iran	48,500	6.9	2,195	3.9
Iraq	44,500	6.3	1,170	2.0
Kuwait	90,000	12.7	985	1.7
Libya	21,100	3.0	1,115	1.9
Neutral Zone	5,400	0.8	420	0.8
Oman	3,500	0.5	420	0.8
Qatar	3,400	0.5	425	0.7
Saudi Arabia	169,000	23.9	4,690	8.1
Others	3,700	0.5	345	0.6
North America	42,800	6.1	11,940	20.1
Latin America	83,300	11.8	6,705	11.9
Western Europe	24,700	3.5	3,800	6.5
Sub-Saharan Africa	20,800	3.0	2,054	3.7
Asia/Pacific	18,600	2.5	3,320	5.8
Eastern Bloc/China	83,800	11.8	15,115	26.5
Total	707,200	100.0	57,800	100.0

Source: British Petroleum, *Statistical Review of World Energy* (B.P., London, 1985).

Of all the oil-producing countries in the region, Iraq is most dependent on international pipelines for exporting its oil. Until recently, almost all of its oil was pumped from Kirkuk to the Mediterranean ports of Tripoli in Lebanon and Baniyas in Syria. However, these lines have been very unreliable because of frequent political disagreements and transit fee disputes. The most serious closures, which can only be understood in the context of the bitter enmity between the Syrian and Iraqi regimes, were between 1976 and 1979 and from 1982. Iraq has done much to lessen its dependence on trans-Syrian pipelines in the last decade, however. In 1976 it opened a strategic pipeline to Al-Faw on the Persian/Arabian Gulf. The following year, oil began flowing through a 986-km line through Turkey to Yumurtalık on the Mediterranean. Iraq has also looked into the possibility of building lines through Kuwait to the Gulf and through Jordan to Aqaba. The most advanced plans, though, are for a line that would link up with and then parallel Petroline to Yanbu on the Red Sea.

A decade of pipeline construction has diversified the Middle East's oil export routes and significantly reduced their vulnerability. Countries like Saudi Arabia and Iraq have considerably greater flexibility, while traditionally vital routes through Syria and Lebanon have lost much of their importance. Planned pipelines, and ones currently under construction, will reinforce these trends.

KEY REFERENCES

McCaslin, J. C. (ed.), *International Petroleum Encyclopaedia* (Pennwell Publishing Company, Tulsa, Oklahoma, 1982).
Oil and Gas Journal and *Petroleum Economist* (various issues).

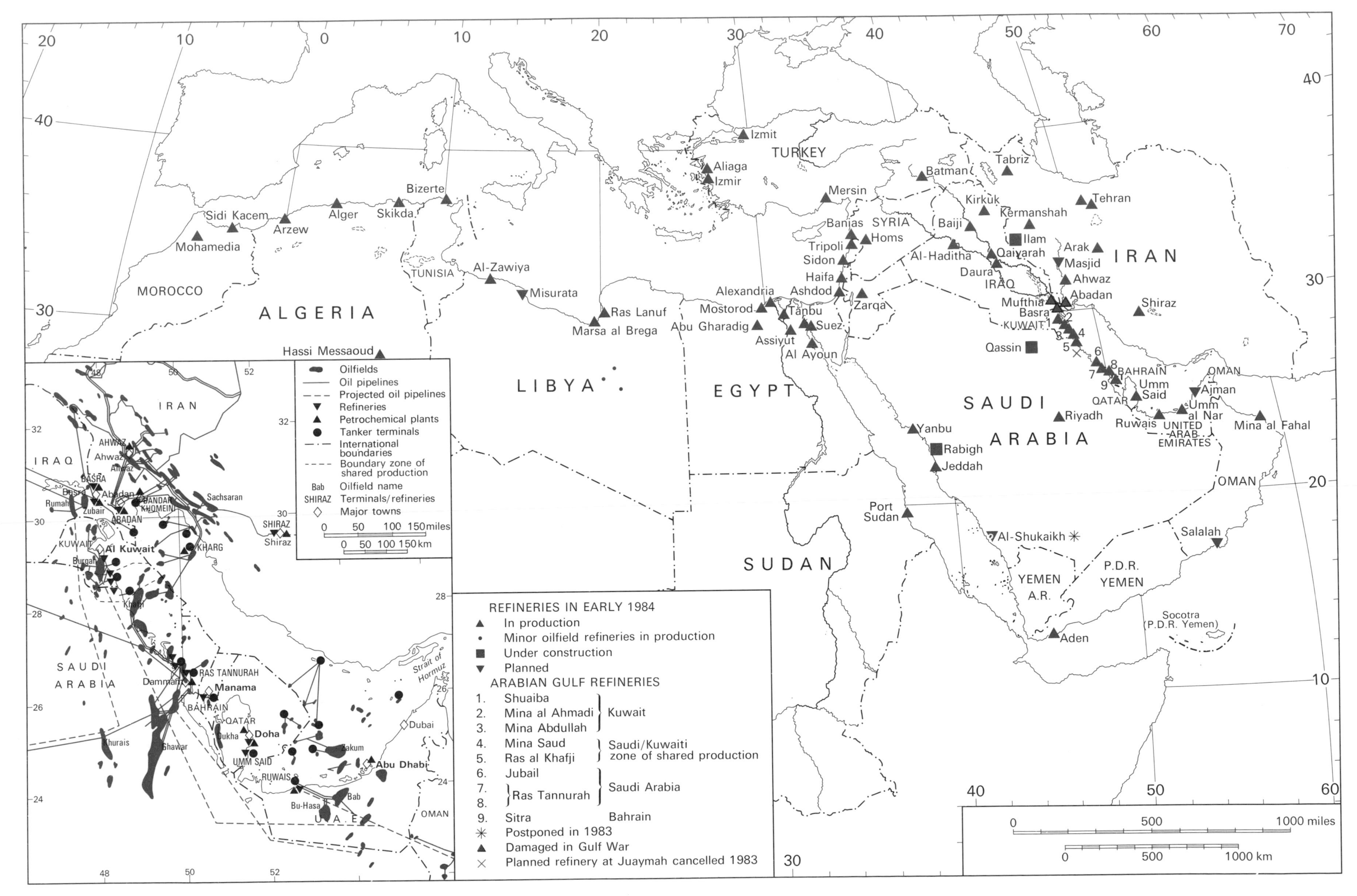

34 Oil refineries

Oil refineries

Alasdair Drysdale

Although the Middle East and North Africa produce one-quarter of the world's oil, they currently refine only 6% of it (see Table 11). Nevertheless, many countries have major petroleum refining and petrochemical industries and have ambitious plans to encourage downstream activities as a means of broadening their industrial bases.

Petroleum refining is by no means new to the region. The first refineries began operations in 1913 at Abadan in Iran and Suez in Egypt shortly after the discovery of oil. However, before World War II Abadan accounted for most of the region's capacity and mainly produced fuel oil for the Royal Navy. After the war, as crude oil production increased and economies expanded, additional refineries were built in Egypt, Iran, Iraq and Saudi Arabia, primarily to refine crude for the growing local market. Even countries with little or no oil of their own, like Syria, Jordan and Lebanon, developed small refining industries. Elsewhere, notably Kuwait, Bahrain and Saudi Arabia, refineries were built with the export market in mind. Since the early 1970s local refining has been boosted by heavy demands for feedstock in the growing petrochemical industry (Map 37). The Gulf states have embarked on ambitious projects to capture some of the forecast increases in the demand for petrochemicals, which have since proved far too high.

Since 1976, the region's refining capacity has increased by over 40% and several additional major projects are scheduled for completion before 1990. In Algeria the Skikda refinery, with a capacity of 300,000 b/d, came on stream in 1983. Libya's 250,000 b/d refinery at Ras Lanuf began operations in 1984. Egypt more than doubled its capacity. Two new facilities opened in Abu Dhabi. Iraq commissioned a refinery at Baiji in 1983 and was due to complete another one in 1985. In Kuwait, the Mina Abdullah refinery is being modernised and output increased. Saudi Arabia, which is already the largest refiner in the region with a capacity early in 1984 of 840,000 b/d, opened or planned to open three more large refineries in 1984–5 at Yanbu, Jubail, and Rabigh with a capacity of over 550,000 b/d. Its huge Ras Tannurah refinery is also being modernised.

All this expansion is causing apprehension in Western Europe and the Far East because it will aggravate the existing surplus capacity of refineries and petrochemical plants. Thus in 1984 only about 75% of the world's refining capacity was utilised. Refineries in the Middle East and North Africa were similarly underutilised and a few projected refinery expansion and construction schemes have been cancelled.

Table 11. *Oil refining capacity*

	Capacity (000 b/d)	Refineries
Algeria	435	4
Abu Dhabi	130	3
Bahrain	250	1
Egypt	369	6
Iran	530	4
Iraq	470	8
Israel	190	2
Jordan	100	1
Kuwait	543	5
Lebanon	52	1
Libya	129	3
Morocco	74	2
Oman	48	1
Qatar	63	1
Saudi Arabia	840	4
Syria	228	2
Tunisia	34	1
Turkey	460	4
Total	4719	54

Source: The Middle East and North Africa, 1984–85 (Europa Publications, London, 1984), and *Oil and Gas Journal*, 31 December 1984.

Refineries, like pipelines, are vulnerable in times of political conflict. Egypt's Suez refineries, for example, were badly damaged in the 1967 Arab–Israeli war. Lebanon's Sidon and Tripoli refineries suffered during that country's civil war and as a result of the feud between Syria and Iraq. Refineries in both Iran and Iraq have sustained heavy damage in the Gulf war. The 610,000 b/d Abadan refinery was virtually destroyed and has been out of operation since 1980. When the war is over, Iran may build a new refinery in a less vulnerable part of Khuzestan rather than rebuild Abadan. Iraq's losses have not been so great, but its 140,000 b/d facility at Basra was badly damaged early in the war and has not been repaired.

Almost all the region's 60 or so oil refineries are located on the coast. Refineries at inland locations tend to be small, and provide largely for a local market. Other refineries are in coastal locations because of their export orientation. Relatively little of the total throughput of refined products is consumed in the area, although local demand for petroleum products is rising. Middle East refineries have an advantage over their competitors in being up to date, and capable of producing high-value light products such as gasoline. About 10% of oil exported from the Middle East and North Africa is now in the form of refined products, whose transportation tends to be more costly than that of crude oil because of the need to use smaller, more specialised ships.

KEY REFERENCES

Exxon Background Series, *Middle East Oil and Gas* (Exxon Corporation, New York, 1985).

International Petroleum Encyclopedia (Pennwell Publishing Company, Tulsa, Oklahoma, 1982).

Oil and Gas Journal and *Petroleum Economist* (various issues).

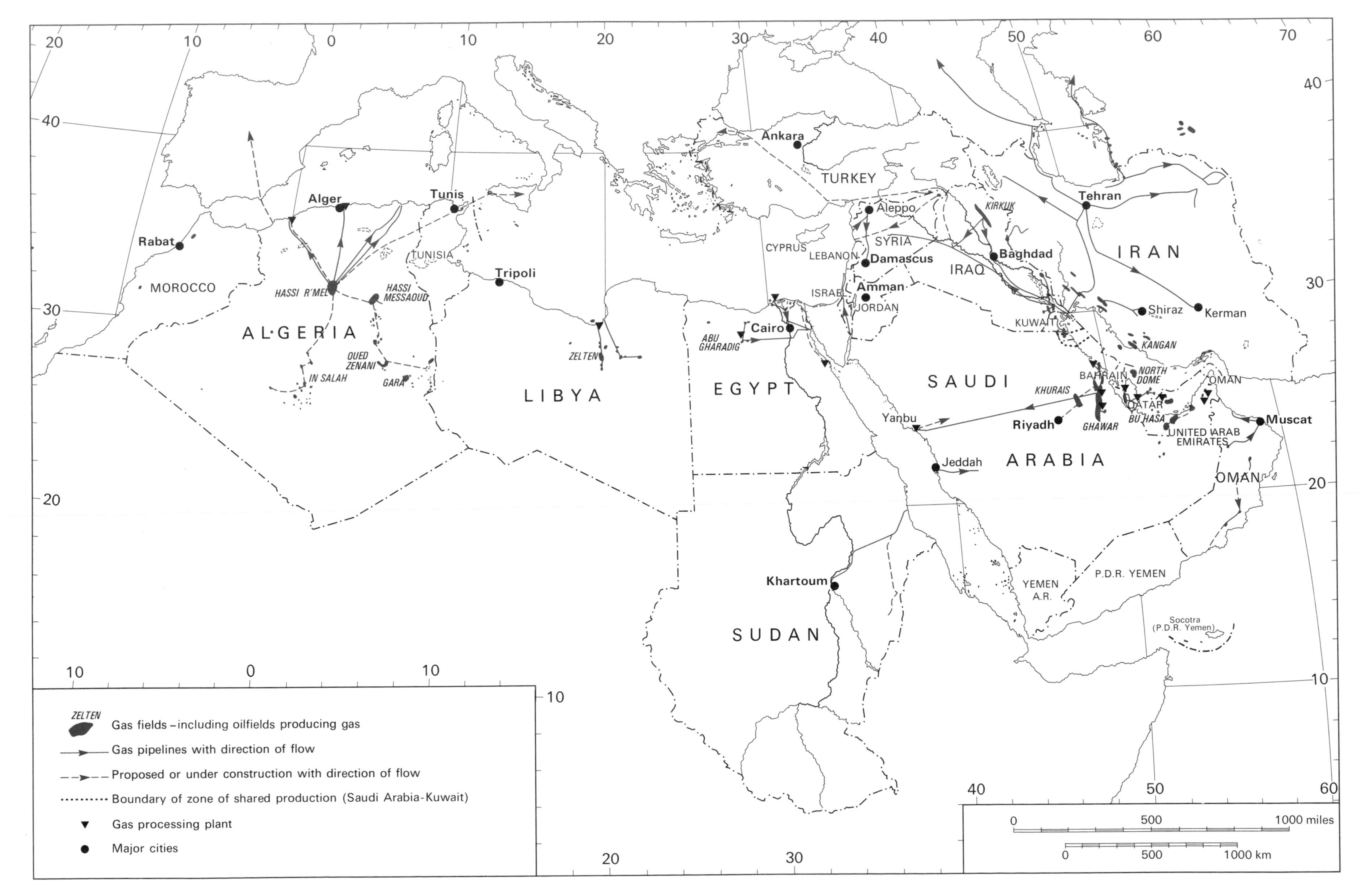

35 Natural gas

Natural gas

Alasdair Drysdale

The Middle East and North Africa have 29.6% of the world's proven natural gas reserves, more than any other region outside the communist world (see Table 12). Despite the close association between oil and gas, countries with most of one are not necessarily those with most of the other. Indeed, 85% of the region's proven gas resources are located in only four countries: Iran, Qatar, Saudi Arabia and Algeria. Iran's reserves alone account for almost half of the region's total and are second only to the Soviet Union's. However, proven reserves significantly understate the actual amount of gas in the region, particularly within certain countries, and have little bearing on how much gas is actually produced or used.

Natural gas may be 'associated' (found in conjunction with crude oil, either in solution or as an overlying gas cap) or 'non-associated' (found on its own). Associated gas, because it is under pressure, helps to drive oil to the surface. This gas, whose production is proportional to the amount of oil produced, may be flared, reinjected into the oil reservoir to maintain pressure, or collected and used as a fuel. By contrast, the production of non-associated gas, because it is found in reservoirs containing only gas, can be calibrated to market conditions. In the Middle East, some 53% of proven gas reserves are associated, although in some countries the share is much greater. An even higher percentage of gas actually produced is associated because most countries prefer to keep their non-associated fields in reserve. This helps to explain why much of the gas reaching the surface in the Middle East is simply burnt off at the well-head: it is the oil that is wanted, not the gas. As recently as the early 1970s probably in excess of 70% of gas was flared because there was little demand for it locally and transporting it to the industrialised consuming nations by pipeline or in liquefied natural gas (L.N.G.) tankers was impractical or prohibitively expensive. Since then, most countries have tried to use their gas more productively to take advantage of rising energy prices, to improve oil recovery, and to fuel ambitious industrialisation programmes. Nevertheless, perhaps one-half of all gas is still flared. In certain countries, the waste is far greater: Iraq, for example, burned off 84% of the gas it produced in 1983 because it has no domestic grid and uses relatively small amounts of its gas industrially. In 1980, when Saudi Arabian oil production peaked at 10 million barrels daily, 38.4 billion cubic metres of associated gas were flared, or slightly less than the total gas consumption of the Middle East or West Germany.

The size and importance of the gas industry varies considerably among the countries of the region. Algeria was one of the first to develop its gas resources and depends more heavily on them economically than any other country in the region. Whereas it has a very low reserve-to-production ratio for oil (27 years), it has vast amounts of gas – the most in Africa. Equally important, much of this gas is non-associated. Today, Algeria is by far the largest gas producer in the region and flares less than 3% of output. About half of the gas is reinjected to increase oil recovery. Much of the remainder is exported as L.N.G. or via the TransMed pipeline to Italy which opened in 1983. However, exports have not lived up to expectations because of competition from North Sea and Soviet gas. Algeria's two L.N.G. plants are operating at only half capacity and several buyers have refused to take as much gas as they originally contracted for.

Iran, with 14% of the world's proven gas reserves and a relatively low reserve-to-production ratio, has one of the most well-developed gas industries in the Middle East. The IGAT 1 pipeline, designed to supply gas to the Soviet Union and northern Iran, began operations in 1970 and plans for a second, much larger parallel line were well advanced at the time of the 1979 revolution. A plant to liquefy gas for Japan was also under construction at the time of the revolution.

Table 12. *Natural gas reserves*

	Reserves (trillion m³)	Share of total (%)
Middle East and North Africa	28.5	29.6
Abu Dhabi	0.6	0.6
Algeria	3.1	3.2
Bahrain	0.2	0.2
Dubai	0.1	0.1
Egypt	0.2	0.2
Iran	13.6	14.1
Iraq	0.8	0.9
Kuwait	0.9	1.0
Libya	0.6	0.6
Qatar	4.2	4.4
Saudi Arabia	3.5	3.6
Others	0.7	0.7
North America	8.2	8.5
Latin America	5.3	5.4
Western Europe	5.8	6.1
Sub-Saharan Africa	1.4	1.5
Asia and Australia	4.6	4.9
Centrally planned economies	42.4	44.0
Total world	96.2	100.0

Source: British Petroleum, *Statistical Review of World Energy* (B.P., London, 1985).

Currently, Iran exports no gas. Although the revolution and the war with Iraq have had a very disruptive effect – 1984 production was less than half the 1979 level – Iran has extremely ambitious (and over-optimistic) plans to expand the domestic grid around the IGAT 1 and 2 lines. By 1990, 183 towns and 400 industrial centres should be connected by 4000 km of spur lines and per capita consumption of gas should approach levels in Europe.

Kuwait and Saudi Arabia illustrate the drawbacks of exploiting associated gas. Kuwait uses its gas in water desalination plants and to produce electricity, ammonia and cement. Because gas production is tied to oil production, which fell dramatically after 1980, there have occasionally been water and power shortages. One way around this would be to import gas when market conditions dictate lower oil output. Kuwait therefore supports a Gulf Co-operation Council plan to build a grid connecting its associated fields with non-associated ones in the lower Gulf, particularly Qatar's enormous offshore North Dome field. When oil production is high, associated gas would flow south; when it is low, non-associated gas would flow north.

Saudi Arabian gas production is subject to even wider fluctuations because of the country's role as chief swing oil producer within O.P.E.C. Between 1981 and 1983, the amount of gas marketed fell from 19.6 billion cubic metres to only 5.4 billion cubic metres. This has made planning domestic consumption problematical. At present most non-flared gas is used to generate electricity, desalinate water and fuel heavy industry. Current development plans project a vast increase in demand as petrochemical, steel, sulphur and many other new industries are established. Accordingly, $20 billion is being spent on a master gas system, which will gather and process virtually all of the kingdom's associated gas. By 1986, some 13,000 km of pipeline will connect four major processing centres.

Despite extensive export and domestic market-oriented infrastructural development and rapid gas-fuelled industrialisation in many countries since the early 1970s, the Middle East and North Africa are, in global terms, not yet major producers or consumers of gas. In 1984, the region produced only 4.1% of the world's non-flared and non-reinjected gas (less than the Netherlands), and consumed about 3% (less than the United Kingdom).

KEY REFERENCES

McCaslin, J. C. (ed.), *International Petroleum Encyclopaedia* (Pennwell Publishing Company, Tulsa, Oklahoma, 1982).

Oil and Gas Journal and *Petroleum Economist* (various issues).

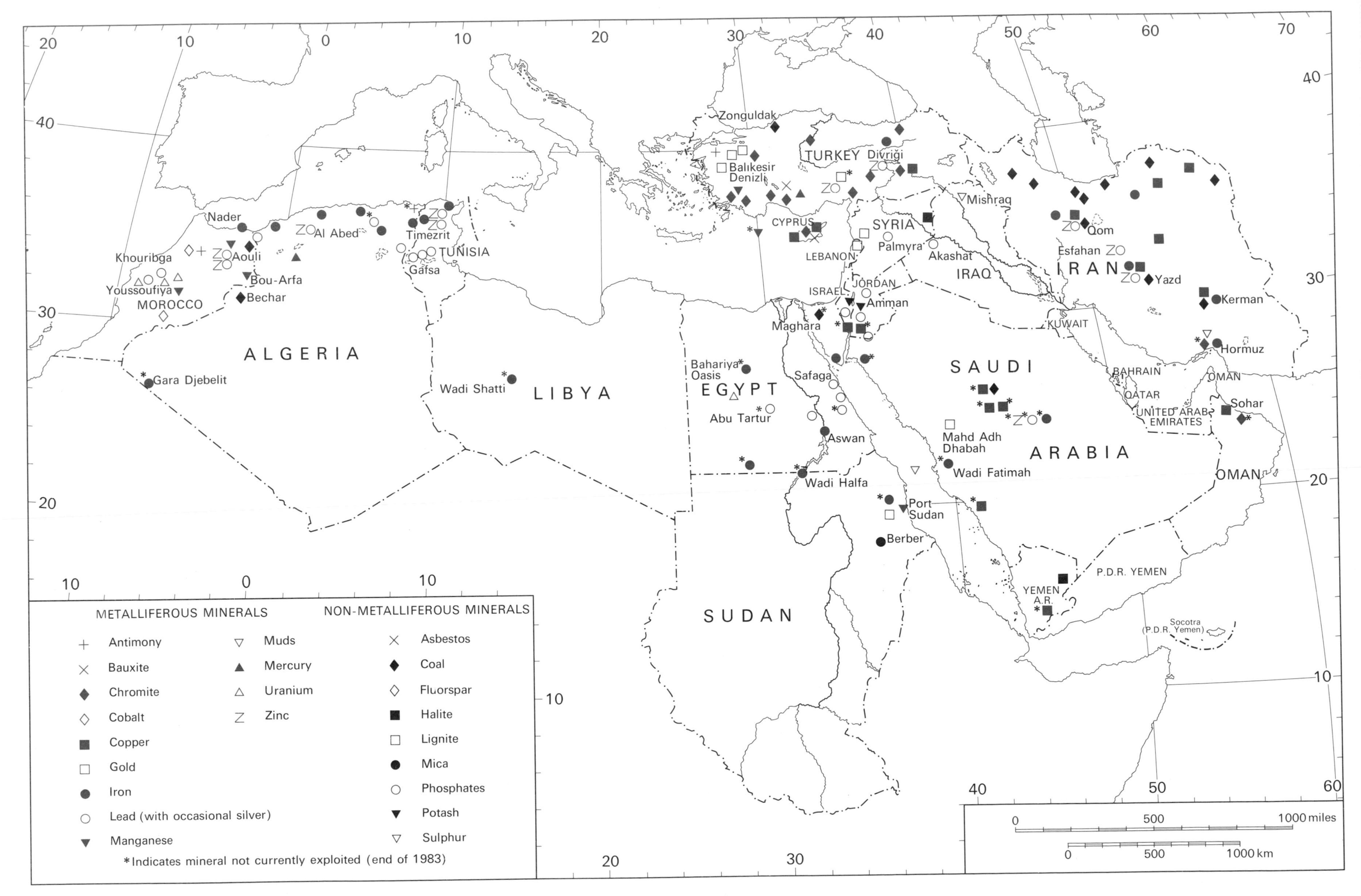

Zonguldak
Balıkesir
Denizli
TURKEY
Divriği
Mishraq
CYPRUS
SYRIA
Palmyra
LEBANON
Akashat
IRAQ
ISRAEL
JORDAN
Amman
Maghara
KUWAIT
Qom
Esfahan
IRAN
Yazd
Kerman
Hormuz
BAHRAIN
QATAR
UNITED ARAB EMIRATES
OMAN
Sohar
SAUDI
ARABIA
Mahd Adh Dhabah
Wadi Fatimah
Nader
Al Abed
Timezrit
TUNISIA
Gafsa
Khouribga
Aouli
Bou-Arfa
Youssoufiya
MOROCCO
Bechar
ALGERIA
Gara Djebelit
Wadi Shatti
LIBYA
Bahariya Oasis
EGYPT
Abu Tartur
Safaga
Aswan
Wadi Halfa
Port Sudan
Berber
SUDAN
YEMEN A.R.
P.D.R. YEMEN
Socotra (P.D.R. Yemen)
METALLIFEROUS MINERALS
Antimony
Bauxite
Chromite
Cobalt
Copper
Gold
Iron
Lead (with occasional silver)
Manganese
Muds
Mercury
Uranium
Zinc
NON-METALLIFEROUS MINERALS
Asbestos
Coal
Fluorspar
Halite
Lignite
Mica
Phosphates
Potash
Sulphur
*Indicates mineral not currently exploited (end of 1983)
0 500 1000 miles
0 500 1000 km

36 Minerals

Minerals

Jonathan Mitchell

The Middle East and North Africa is not a region particularly noted for mineral production. Map 36 shows that a considerable variety of minerals are found, however, and many are commercially exploited. But to put the situation in perspective, only Algeria and Morocco produce more than 5% of the world total for specific major minerals (Table 13). Furthermore, only in the cases of Morocco and Jordan are minerals the major national export items by value, and in all cases minerals account for less than 3% of G.D.P.

Most established mineral workings are in the Alpine fold regions of the Maghreb, Turkey and Iran. The exception are phosphate deposits which occur in sediments overlying older rocks to the south of the Alpine systems. New mineral deposits are being located in the established areas. The Sar Cheshmeh copper deposits near Kerman in Iran are a good example, but most explorative work is now concentrated in the vast areas of the Pre-Cambrian shield of Saharan North Africa and Arabia. A number of potentially valuable mineral deposits have been located. Algeria, Libya, Egypt and Saudi Arabia have all discovered deposits of iron ore in the last decade, and Saudi Arabia will be producing gold by 1987. However, many of the deposits in the Pre-Cambrian shield areas are in very isolated locations. Commercial exploitation frequently involves considerable investment in road and rail transport. Both Algeria and Libya are planning rail links to their respective major iron ore deposits (see Map 42), and the exploitation of copper at Sar Cheshmeh, albeit in an 'established area', has required new roads and a complete new town. This isolation means that often only the largest or most valuable deposits can be considered commercial propositions.

Another area of exploration is the ocean floor. Research is most advanced on the Red Sea bed. In 1974 the Saudi Arabians and the Sudanese set up the Red Sea Joint Commission to co-ordinate research and exploitation. Commercial extraction of metalliferous muds is now set to commence in 1988. Metalliferous muds have also been located off the southwest coast of Turkey.

Current explorative activity in the new areas reflects a desire to diversify national economic bases against the time when oil runs out rather than simply to increase national incomes as is the case in the oil deficient states. It is significant, though, that few of the countries with known appreciable mineral resources have the kind of minerals desirable for a heavy industrial base. Few have both metal minerals and coal, for example, though recent discoveries now suggest that Saudi Arabia may be particularly well endowed in this respect. Whether the quantity and type of minerals will be sufficient and the world economic climate right for economic diversification to be successful remains to be seen.

Table 13. *Major mineral production*

Mineral	Producer	Output (000 tonnes)	World ranking	Percentage of world production
Antimony (1982)	Turkey	1,237	7	2.4
	Morocco	845	11	1.6
Chromite (1980)	Turkey	196	5	4.6
	Iran	39	11	0.9
Copper (1982)	Turkey	1,491	22	0.3
Iron ore (1981)	Algeria	1,807	20	0.4
	Turkey	1,491	22	0.3
Lead (1982)	Morocco	105	9	2.9
Lignite/brown coal (1981)	Turkey	13,500	15	1.3
Manganese ore (1980)	Turkey	499	7	3.8
Mercury (1982)	Algeria	600	5	9.1
	Turkey	229	6	3.5
Phosphate rock (1980)	Morocco	18,824	3	13.9
	Tunisia	4,502	5	3.3
	Jordan	3,911	6	2.9
	Israel	2,307	9	1.7
	Syria	1,319	13	1.0
Silver (1982)	Morocco	52	19	0.5

Note: Metal ore output is given in metal content, phosphates as P_2O_3 content.
Sources: Geographical Digest 1984 (George Philip and Son, London, annual), and U.N., *1981 Statistical Yearbook* (United Nations, New York, 1983).

Table 14. *Major mineral developments*

Country	Developments
Algeria	2000 m tonne medium grade iron ore deposit at Gara Djebelit was due to begin production in 1985–6.
Egypt	Medium grade iron ore in commercial quantities located at Bahariya Oasis.
	1000 m tonne phosphate deposit located at Abu Tartur.
	Re-opening of Maghara coal mine in Sinai planned.
Iraq	Akashat phosphate mine opened in 1981 – projected annual output of 3.4 m tonnes.
Iran	Mining of 800 m tonnes of copper ore at Sar Cheshmeh commenced 1982.
Libya	700 m tonne iron ore deposit at Wadi Shatti – production imminent.
Morocco	Traces of uranium found in the High Atlas.
Oman	Exports of copper from 50 m tonne deposits near Sohar in the northwest began in 1983.
Saudi Arabia	1.2 m tonne gold ore deposit at Mahd adh Dhabah – mine inaugurated 1983.
	Commercially viable coal deposits located.
	Metalliferous muds under investigation on Red Sea bed – exploitation set to commence in 1988.
Syria	Feasibility studies underway on exploitation of 530 m tonne iron ore deposits.
Turkey	400,000 tonne manganese deposit located near Denizli.
Yemen A.R./P.D.R. Yemen	Extensive surveys carried out – possible joint exploitation of copper ore at Beida.

Sources: The Middle East and North Africa 1983–84 (Europa Publications, London, 1983); *The Middle East and North Africa 1984–85* (Europa Publications, London, 1984).

KEY REFERENCES

Beaumont, P., Blake, G. H. and Wagstaff, J. M., *The Middle East: A Geographical Study* (Wiley, Chichester, 1976).

Mann-Borgese, E. and Ginsberg, N., *Ocean Yearbook 3* (University of Chicago Press, Chicago, 1982), pp. 77–104.

The Middle East and North Africa 1983–84 (Europa Publications, London, 1983).

The Middle East and North Africa 1984–85 (Europa Publications, London, 1984).

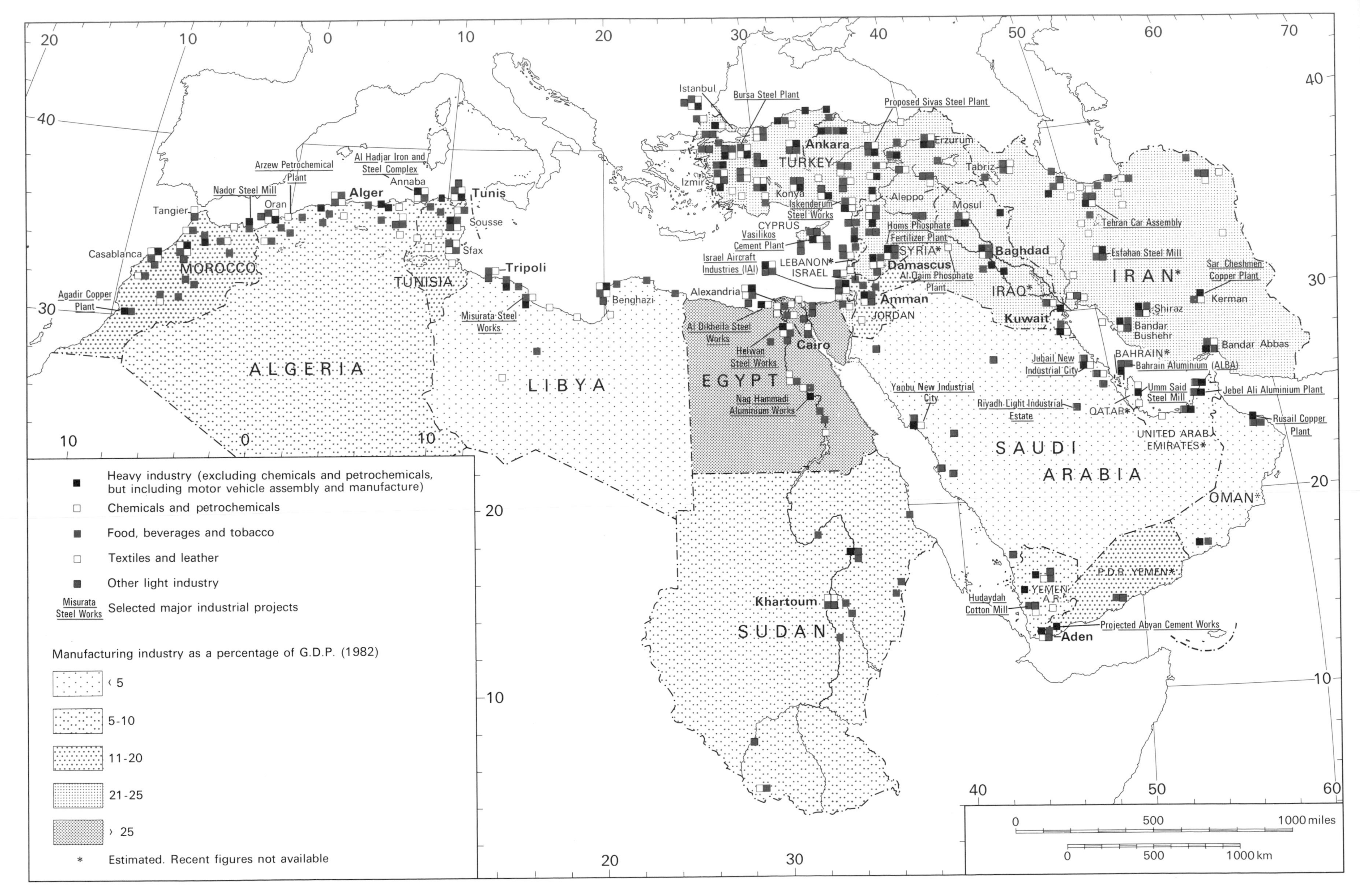

37 Industrial location

Industrial location

Rodney Wilson

In the Middle East, as elsewhere, both economic and political factors play a part in determining industrial location. Economic factors include the availability of industrial supplies and market access, these in turn being affected by the adequacy of transportation facilities. Political factors may include the desire to develop certain regions to favour selected political constituents. Where industry is nationalised, the state will of course make the decision regarding the location of plant, but even when private capital is involved, an industrial permit system is often used as the main control for regional policy.

Much of the modern heavy industry in the Middle East is resource related, either processing crude materials or using them as a source of energy. The best example of the former is the refining of crude oil into petroleum products. Instances of the latter include petrochemicals, aluminium smelting and steel making by direct reduction techniques, all of which use gas as the energy feedstock. Not surprisingly, there is a major concentration of such industries near the oil and gas fields of the Gulf, the largest single centre being at the new industrial city of Jubail in Saudi Arabia. The main products include ethylene, polyethylene, methanol and urea for use in the fertiliser industry. The Jubail plants are at coastal sites to facilitate the shipping of the petrochemical produce, most of which is marketed in the Far East and the European Community.

New heavy industries in the Middle East are usually located on new industrial sites where there is plenty of land for expansion rather than near existing centres of population. The industries are capital intensive and employ relatively small numbers of people once the initial construction phase is completed. Most of the employees are expatriates, who often prefer to live in newly erected housing in company compounds rather than in established towns. Although an attempt is being made to employ more Gulf citizens in the new petrochemical plants, most of the skilled and semi-skilled workers came from South and Southeast Asia, more specialised technicians from Europe, North America and Japan on short-term contracts.

Saudi Arabia's planners hope that both Jubail and Yanbu will act as growth poles for regional development. Further downstream investment is envisaged, mainly by private entrepreneurs, in plants which can use the basic petrochemical products. The Ministry of Industry is having some success with this policy, thanks partly to state financing on concessionary terms for new projects through the Saudi Industrial Development Fund. New industries manufacturing a range of plastic products are being established, mainly for the Saudi Arabian and Gulf regional market. The scale of production is limited, however, and it will be the early twenty-first century at the soonest before Jubail and Yanbu can be regarded as fully integrated industrial complexes.

Outside the Gulf area, Algeria is the only major crude oil producing country to have established significant downstream industries. Its petrochemical plant at Arzew is the largest in the Mediterranean, the products being mainly shipped to France and other European Community countries. Algeria and Kuwait are the only two O.P.E.C. states to export a large proportion of their oil in refined form rather than as crude. As Algeria is also a major exporter of liquefied gas, which is worth almost as much as its refined product exports, the country's energy exports are more diversified than those of any other O.P.E.C. state.

In the more populous Middle Eastern states most of the industries have been set up to serve domestic market requirements for consumer durable and non-durable goods. As a consequence these industries tend to be located near the major concentrations of population. As the urban middle classes are the major purchasers of consumer durables such as cars, televisions, refrigerators and other household electrical goods, such industries are typically situated in the major urban centres, often in the country's capital city. Over three-quarters of the modern manufacturing capacity of Egypt, for example, is located in the greater Cairo area, the industrial suburb of Helwan having the greatest single concentration, including the Middle East's largest steel plant. Similarly, in Iran around 70% of modern manufacturing industry is located around Tehran, including industries such as car assembly. The textile industry is more dispersed, however, with Esfahan and Tabriz being major centres for spinning and weaving as well as Tehran. Much of Iran's traditional carpet weaving industry is regionally dispersed.

Consumer goods industries are usually much more 'footloose' than heavy manufacturing in the sense of being less dependent on sources of raw materials as far as location is concerned. Encouragement can be given through capital grants or subsidies for such industries to establish plants in backward regions with few employment opportunities outside agriculture and serious underemployment problems. However, few Middle Eastern governments have embarked on any rigorous policies for the dispersal of industry.

Fortunately in some Middle Eastern countries the capital city acts as less of a centre of gravity than in others. This is particularly the case where a new location for the capital has been chosen as a result of political factors. In Turkey, for example, Istanbul remains the largest city and major commercial centre, even though the capital was moved to Ankara with the establishment of the Turkish Republic. Hence both cities have proved attractive locations for light industries, as well as the important Mediterranean port of Izmir. In Saudi Arabia, when the capital was moved to Riyadh this acted as an important stimulus to the industrial development of the interior of the kingdom, even though Jeddah remained the largest city, and the main centre for commerce.

There are, of course, economic advantages in industrial concentration in a particular locality. Common services can often be shared by several industries. Close linkages may exist if one plant's output constitutes another industry's input. It is this type of external economies that results in some regions thriving as industrial economies, while others are neglected by manufacturing. Apart from the cases of Egypt, Turkey and Iran, however, the manufacturing bases of the Middle Eastern states are too small for such economies to be realised to any significant extent. Even in Egypt, Turkey and Iran, many consumer durable industries are mere assembly operations, which use imported components. In these circumstances the linkages with other local industries are minimal, and the argument for industrial concentration is weak.

Egypt has become the most industrialised country in the Middle East, with manufacturing accounting for over one-quarter of gross domestic product. The share of manufacturing in gross domestic product in countries such as Iran, Iraq and Algeria is reduced because of the significance of oil in these economies. Nevertheless, even when this is taken into account, the extent of Egypt's manufacturing base is wider, reflecting its longer history of modern industrialisation. Both Egypt and Turkey have over six decades of experience in establishing modern manufacturing. In contrast, in the Gulf states and Saudi Arabia modern manufacturing was only introduced in the 1960s, and most industries have been established for less than a decade.

The existing pattern of manufacturing is likely to prevail into the twenty-first century. The Gulf seems likely to emerge as a major global centre for resource-related, predominantly export-oriented, industries. Its share of world petrochemical production may rise from the 6% recorded in 1985 to over 20% by the turn of the century. Nevertheless, the Gulf region will only diversify slowly beyond basic petrochemicals such as ethylene and methanol. In contrast, manufacturing diversification seems likely to advance faster in the more populous Middle Eastern states, as consumer demand increases, but foreign exchange constraints preclude a high level of imports.

KEY REFERENCES

Clarke, J. I. and Bowen-Jones, H., *Change and Development in the Middle East* (Methuen, London, 1981).

Gueciouer, Adda (ed.), *The Problems of Arab Economic Development and Integration* (Westview Press, Boulder, Colorado, 1984).

Turner, Louis and Bedore, James, *Middle East Industrialisation* (Saxon House, Farnborough, 1979).

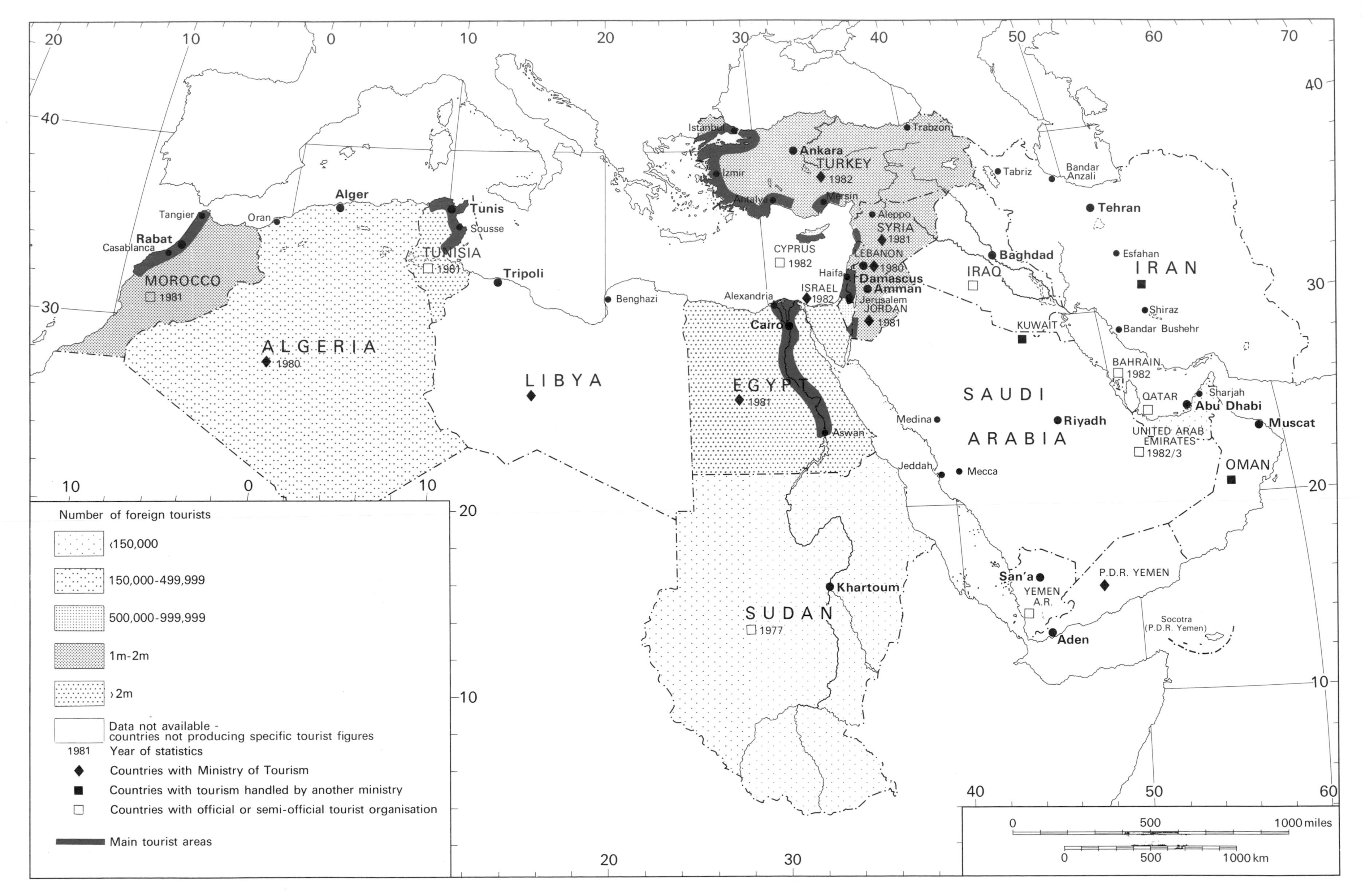

38 International tourism

International tourism

Gerald Blake

By world standards, the Middle East and North Africa is not a major tourist region. Although there are difficulties in collecting accurate statistics, it is probable that no more than 15 million international tourists visit the region each year, which in 1983 represented only about 5% of the 287 million world tourists. Several European states receive twice as many tourists annually as the whole of the Middle East and North Africa. Nevertheless, for a number of the states of the region tourist revenues contribute substantially to G.D.P. and are an important feature of government economic development strategies (Cyprus 13.8%, Tunisia 7.6%, Israel 4.1%, Morocco 3.9%, Egypt 1.2%, Algeria 1.1%, Syria 1.0% in 1983). Tourism may also offer an opportunity for employment and development in local areas with limited alternatives in the agricultural or industrial sectors.

The tourist potential of the region is the product of favourable climate, historic heritage and global location. The Maghreb in particular is only a few hours' flying time away from Western Europe, the world's greatest reservoir of international tourists. The whole Middle East has good communications by sea and air with Europe, and direct access overland is possible via Turkey (Maps 43, 44, 45). The region has potential for both 'up-market' tourists and 'down-market' tourists, or the high-spending minority and the lower-spending masses. In addition to being endowed with long coastlines and abundant sunshine, several countries have high mountain ranges which may be snow-capped in winter and spring. An added attraction is the rich cultural life of the region expressed in architecture, folklore and food. The relative tourist potential of states, and their ability to absorb large numbers of visitors, varies greatly from those with high potential (Morocco, Algeria, Tunisia, Egypt, Turkey) to those with very modest potential (Kuwait, Bahrain, Qatar, United Arab Emirates). Generally, however, government policy is more important than inherent quality in determining the number of visitors. Several states, particularly the more conservative oil-rich states, are frankly not interested in developing a tourist industry, or positively discourage tourism. Ironically, these same states are often heavily dependent upon large numbers of foreign labourers and technical experts. Other states, such as Lebanon, Iraq, and Iran, have been keen to encourage foreign visitors in the past but receive relatively few today because of political turmoil and physical violence. Tourism is notoriously susceptible to political upheavals, and the Middle East and North Africa have frequently lost tourist visitors and revenues because of adverse publicity or closed frontiers in times of crisis.

The map illustrates, as far as possible, the pattern of international tourist arrivals in the early 1980s. All the top tourist countries in the region are located on the fringes of the Mediterranean Sea. Tunisia and Egypt, followed by Morocco and Turkey, are the leading tourist centres. Israel, which attracts visitors to Biblical sites in large numbers and a steady stream of Jewish visitors, has the most carefully planned and packaged tourist industry. Large numbers of Arab tourists go to Egypt, Jordan, Iraq and Syria, although Egypt's Arab tourist numbers declined after the peace agreement with Israel in 1979. Turkey's tourist potential is at last being realised, and that country is now a popular destination for American, European and Middle Eastern visitors. Tourist numbers could rise more sharply in Turkey than in any other state in the region in the next few years. Syrian tourism might also develop considerably if political events allow it to do so. Turkish ports, particularly Istanbul, are popular with cruise ships plying the Black Sea and Aegean in the summer months. Cruise passengers provide a significant source of revenue in a number of Mediterranean ports, including Tangier, Tunis, Alexandria and Haifa. The map does not include the large annual influx of pilgrims into Saudi Arabia to perform the Haj, nor does it show domestic tourism. For many generations, the rich families of the Middle East and North Africa have enjoyed summer vacations in the cooler parts of the region, on the coast or in the mountains. Lebanon traditionally hosted large numbers of such seasonal visitors. More recently, winter ski resorts have appeared in Morocco, Algeria, Lebanon and Turkey, largely for the benefit of local upper-class holidaymakers.

Table 15. *Bed places in hotels and other tourist establishments*

Country	Hotels	Other establishments
Algeria	26,663	
Cyprus	22,917	538
Egypt	43,762	
Israel	50,145	17,066
Jordan	10,229	
Morocco	58,044	43,444
Syria	34,000	
Tunisia	62,370	12,582
Turkey	44,069	14,000

Source: World Tourism Organisation, *World Travel and Tourism Statistics*, vol. 37, *1982–83* (World Tourism Organisation, Madrid, 1984), Table D.

The arguments for and against tourism in developing economies such as those of the Middle East and North Africa are vigorously debated. Tourism can generate jobs and earn foreign exchange while making modest demands upon scarce water and land. On the other hand, it reinforces dependency and may involve high social and environmental costs to the host country. The tension between the package-tourist business and traditional society has been especially painful, for example, in Morocco and Tunisia. Some forms of tourism may also reward foreign investors and entrepreneurs too highly, bringing only modest returns to the local economy. The building of hotels and tourist complexes had undoubtedly contributed to the region's construction boom in recent years. Some of the world's finest hotels have been built in the Gulf region, largely to serve the needs of visiting businessmen. In several major cities such as Cairo, the demand for good hotel accommodation still exceeds the availability of beds. Inadequate infrastructure, particularly roads, also remains a deterrent to the growth of tourism in several parts of the region such as the Yemen Arab Republic and Sudan. Another problem is that Middle East and North African boundaries are often troublesome to cross, and not infrequently closed because of political differences between neighbours. As a whole, the region's international tourist potential is thus underdeveloped. A crude indication of the relative stage of development can be gleaned from numbers of hotel beds (Table 15).

For some states, tourism represents economic opportunity which current development plans intend to exploit, while for others the neglect is deliberate. The demand for access to Middle East and North African locations by international tourists is bound to increase. Besides European visitors, larger numbers from the United States and Japan can be expected. With careful planning and strict environmental controls, tourism could probably benefit the region far more than at present. To maximise the potential of states away from the Mediterranean fringe where 'up-market' visitors offer the best prospects, more regional co-operation may be necessary, while political stability is an essential pre-requisite. As the region becomes a popular destination for tourists, the oil-rich states in particular are an increasing source of very welcome 'up-market' Arab visitors in Western Europe and elsewhere.

KEY REFERENCES

Britannica Book of the Year (Encyclopaedia Britannica, Chicago, 1984).
World Tourism Organisation *World Travel and Tourism Statistics*, vol. 37, *1982–83* (World Tourism Organisation, Madrid, 1984).

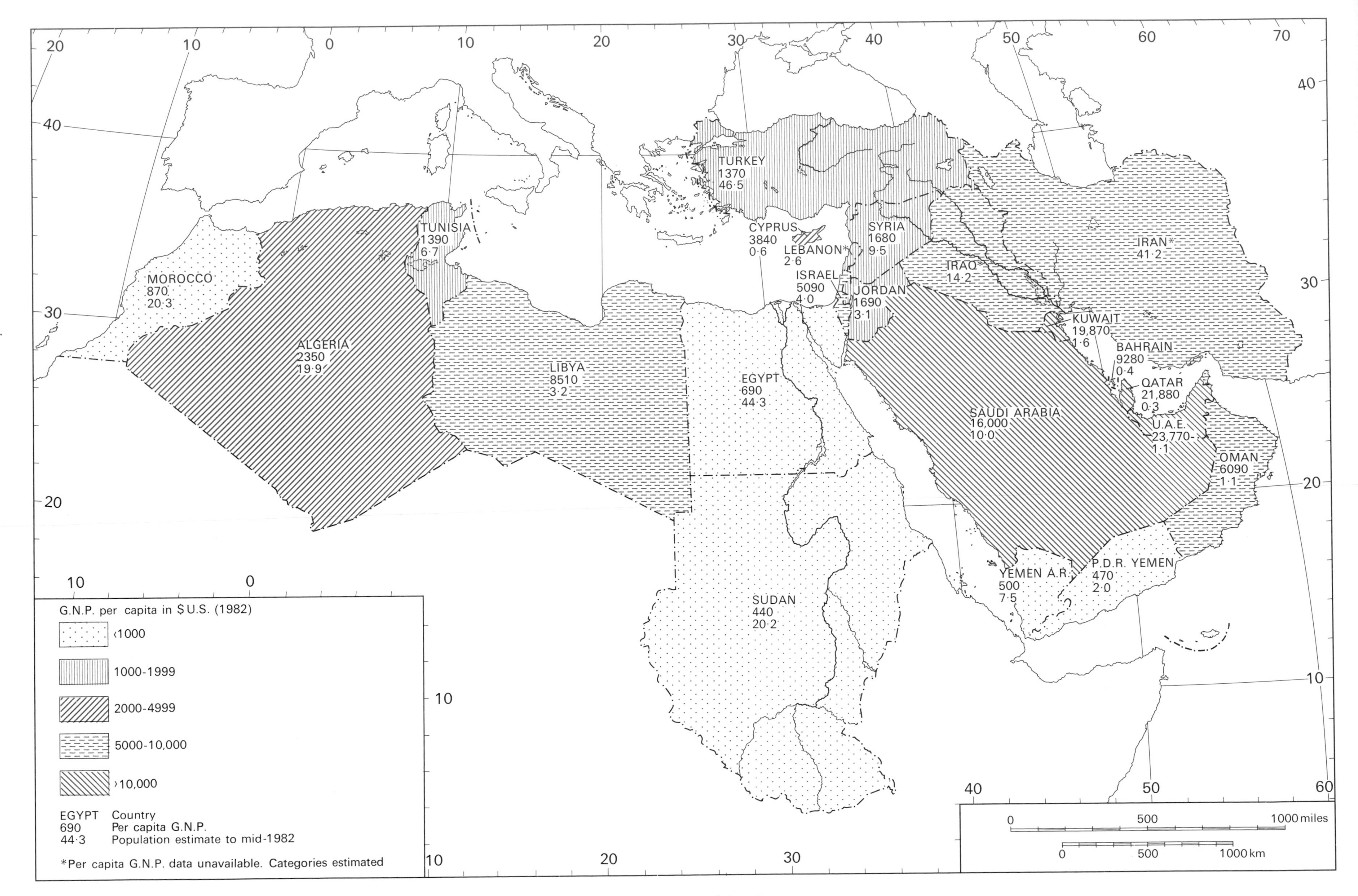

39 Gross National Product

Gross national product

Rodney Wilson

Per capita gross national product (G.N.P.) is the most widely used indicator of a country's level of development. Countries with per capita G.N.P.s of less than U.S. $500 per annum such as the Sudan or the Yemens are classified as low income, and are considered to be amongst the least developed countries in the world. Those countries with per capita incomes of up to $1700, such as Jordan, Syria and Tunisia, are classified as middle income. These three countries all had per capita G.N.P. in excess of $1390 in 1982. The middle income band is wide, however, and also includes countries such as Egypt, whose per capita G.N.P. was only $690 in 1982.

Starvation is rare in middle-income countries, and most people have adequate diets, although grains and vegetables are the major items consumed rather than meat or fish. At the upper end of the middle-income band, as in Syria or Jordan, most households own consumer durables such as televisions, but car ownership is much less widespread. Houses will usually be brick or stone built, with concrete or wooden floors, in contrast to the low-income countries where houses are of a less permanent nature, being constructed of mud bricks or straw, the floors usually being bare earth which cannot be kept clean.

The major oil-exporting countries are in a category of their own, as states such as Kuwait or the United Arab Emirates enjoy the highest per capita G.N.P. in the world at over $23,770 in 1982. This is greater even than that of the United States at $13,160 or Switzerland at $17,010, the latter being the most affluent country in Europe. Even a more populous oil-exporting country like Saudi Arabia recorded a per capita G.N.P. of $16,000 in 1982, over 38% greater than that of the United Kingdom. The per capita G.N.P. figures for the Gulf states and Libya are, however, a striking instance of the inadequacy of this indicator as a measure of living standards. Although the major oil-exporting countries are prosperous by Third World standards, and food and housing conditions for local citizens are generally adequate, living standards remain well below those enjoyed by the majority in Western Europe, Japan or North America. Per capita G.N.P. merely reflects the level of oil production. A high level of oil production increases current national and personal income, but it reduces future potential income, as oil reserves are depleted more rapidly.

Living standards may of course reflect national wealth as well as national income. Housing which was built centuries ago may still be used, and may contribute to a nation's living standard. The countries of Western Europe have a rich inheritance of accumulated capital in the form of housing, railways and roads which reduces their need for income to finance new investment. Middle Eastern countries such as Egypt also have a considerable inheritance of accumulated capital. In rural Egypt, irrigation canals built centuries ago are still in use today. The railway system constructed in the last century still serves the country well.

In contrast, in the historically nomadic societies of the Gulf and Libya there was little capital accumulation. Wealth was traditionally in the form of animals, and not usually in buildings or irrigated land. Such countries therefore need a higher level of income merely to build up their capital stock. Comparisons of per capita G.N.P. figures with more established societies such as those of Egypt or Syria may therefore be misleading. It would be wrong to conclude, for example, that Libya was 12 times richer than Egypt simply because that reflected the per capita G.N.P. differential.

A nation's wealth also reflects the quality of its human capital as well as its physical capital. Here again, the contrasts in the Middle East are striking. Countries such as Egypt or Iran have been centres of learning and scholarship for centuries, which encompassed a wide range of disciplines. Both countries have substantial minorities who have benefited from secondary education, and the number of university graduates in each country exceeds half a million. In the oil states of the Gulf and Libya, modern education systems have been established only recently. It is only the young who are benefiting from education, and most of the older generation, even those in senior positions, have enjoyed little formal education. Yet none of these factors is taken into account in the G.N.P. per capita estimates.

Per capita G.N.P. is usually thought to be a better indicator of living standards than gross domestic product (G.D.P.). The latter only takes account of the value of domestic production, while the former includes income from abroad. In the Middle East the latter is significant for both the major oil exporters and the poorer, more populous states. The remittances which Egyptian workers earn in the Gulf or Turkish workers earn in West Germany are taken into account in the G.N.P. figure. For the oil-exporting countries, the income on overseas investments is an important element in G.N.P., but not G.D.P. Kuwait earned more from its overseas investments in the West in 1981 than it did from oil, and one-third of Saudi Arabian income was accounted for by the interest, profits and dividends on foreign financial assets.

Per capita G.N.P. for different countries has to be denominated in terms of a common unit of account if international comparisons are to be made. The figures cited on the map are in U.S. dollars, the usual conversion, because of its widespread recognition and use. If a currency depreciates, however, this lowers the value of its G.N.P. when measured in dollars, and appreciation increases its value. Exchange rates have generally been stable in the Middle East but, as Turkey and Israel devalue their currencies several times a year, the choice of a conversion rate poses a problem. When countries have multiple exchange rates, as is the case with Egypt, the conversion factor is also a matter for debate. This type of issue means that G.N.P. comparisons are far from being precise, and must be interpreted with caution.

KEY REFERENCES

Wilson, Rodney, *The Arab World: An International Statistical Directory* (Wheatsheaf Books, Brighton and Westview Press, Boulder, Colorado, 1984).

Wilson, Rodney, *The Economies of the Middle East* (Macmillan, London and Holmes and Meier, New York, 1979).

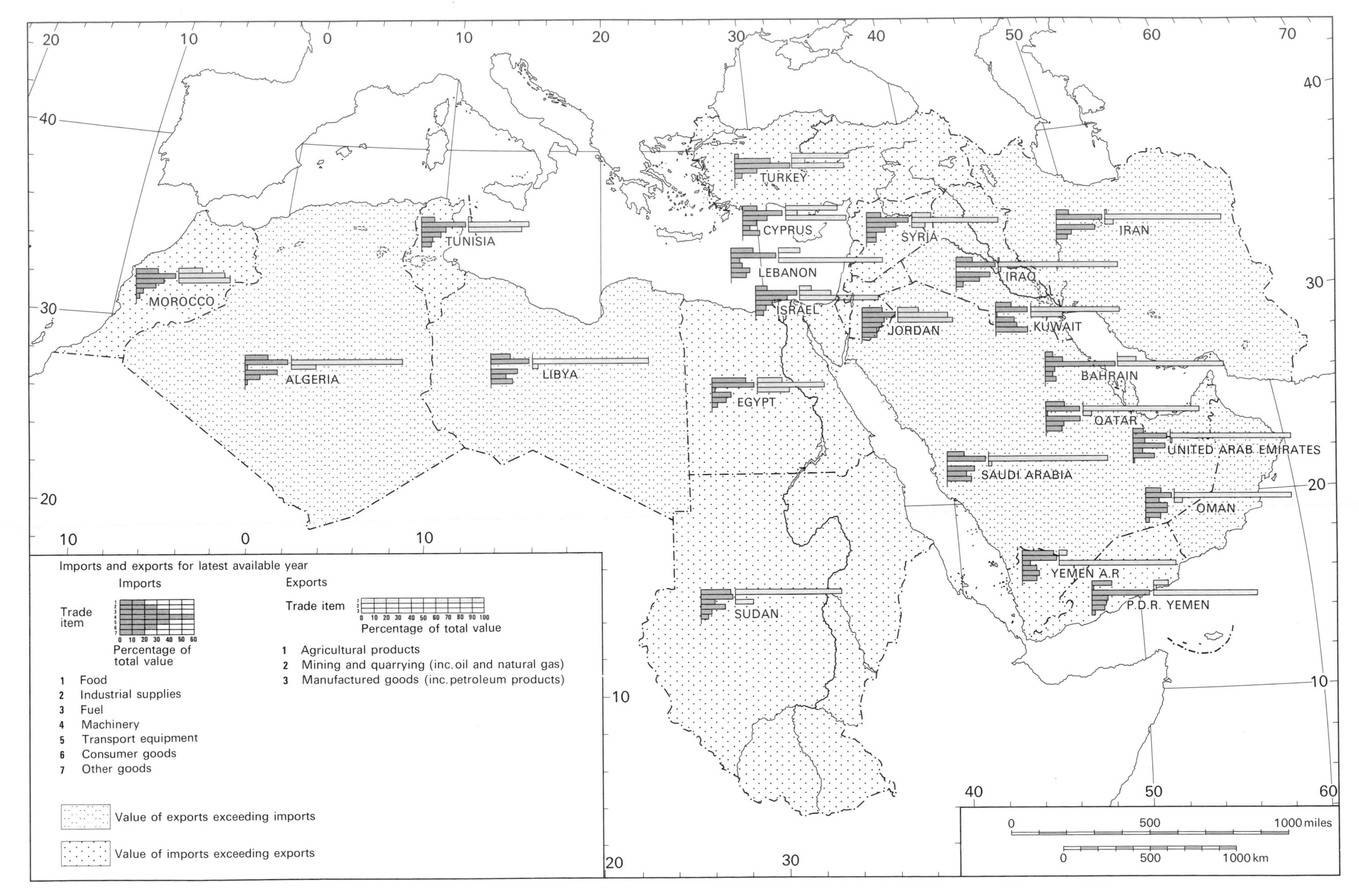

40 Trade, exports and imports

Trade, exports and imports

Rodney Wilson

From a trading point of view, the economies of the Middle East can be divided into two categories: the major oil exporters which are members of the Organisation of Petroleum Exporting Countries (O.P.E.C.), and the non-O.P.E.C. states, producing oil only in limited quantities, or not at all. The states which border the Gulf are of course the major oil exporters, together with Algeria and Libya. In most of these states, oil accounts for over 90% of export earnings. In contrast, in the non-O.P.E.C. states exports are much more diversified, and include both agricultural produce and industrial goods.

Historically, export diversification has been the exception rather than the rule. Egypt and the Sudan were, for example, dependent upon raw cotton exports for virtually all their visible overseas earnings until the 1960s. Gulf states such as Bahrain and Qatar were almost completely reliant on exports of pearls down to the 1930s. In Iraq, exports of dates predominated, and for Iran overseas sales of handwoven carpets were the major source of foreign exchange before the advent of oil.

The problems of such export concentration which were evident in the past still remain today. Countries are dependent on world markets over which they can exercise little or no control for the pricing of their major exports. Of course, O.P.E.C. had a major influence on petroleum price developments in 1973–4 and to a lesser extent in 1979, but historically oil, like other primary commodities, has tended to decline in price. This is particularly the case in relation to prices of manufactured goods. As the Middle Eastern states mainly export primary commodities and import manufactured goods there has been a long-term trend for export prices to decline vis-à-vis import prices. In other words their terms of trade have deteriorated. This has been due largely to demand factors, as the income elasticity of demand is higher for manufactured goods than for primary produce. Hence as world income rises, the demand for manufactured goods increases more rapidly than that for primary commodities, and a growing proportion of expenditure is on the former.

An additional difficulty is the instability of prices of primary produce in world markets, and the consequences for variability of export earnings. Jordan, for example, is heavily dependent on the export of phosphates, but prices have sometimes doubled and halved again within a very short time span, as in 1976–7. Such fluctuating export proceeds make it difficult to plan ahead, as any development programme needs a regular supply of foreign exchange to meet import payments. If a project is started and then curtailed because of temporary payment difficulties, then there is much waste of resources and effort.

One aim of O.P.E.C. has been to stabilise prices by acting as a cartel. It has only been able to achieve some measure of stability by cutting supply, and hence export proceeds have declined. In Saudi Arabia oil earnings fell by more than two-thirds over the 1981–5 period. Hence price stability has only been achieved through output instability. Price predictability implies quantity unpredictability.

As a result of the long-term tendency for the terms of trade of primary producers to decline, and problems over short-term export price instability, Middle Eastern states have attempted to diversify their range of exports. Such structural adjustments take time, however, and can be economically painful. Egypt, for example, has attempted to diversity out of raw cotton exports into the export of cotton yarn, cotton cloth and even made-up clothing. Tunisia has developed an export-oriented clothing industry by encouraging French companies to establish local subsidiaries and to participate in joint-venture enterprises. Turkey has tried to penetrate the European Community market for ready-to-wear garments. In all cases these efforts have been frustrated. Despite having Association Agreements with the European Community, restrictions have been imposed against Middle Eastern textile exports. Quotas have been introduced under the Multi-Fibre Agreement, as European governments have been concerned with the problems facing their own ailing textile industries.

Similar frustrations are being experienced by Middle Eastern oil exporters. European Community countries prefer to purchase crude oil rather than refined products from the Middle East. There is much excess refining capacity in Western Europe, and the import of refined products only accelerates refinery closures. The European Community views Middle Eastern petrochemicals as an even greater threat. Tariffs have been applied to the petrochemical produce exported from the new complex at Jubail on the Gulf by the Saudi Arabian Basic Industries Corporation (S.A.B.I.C.). The oil producers naturally wish to see more value added carried out near the oilfields, and hence are diversifying downstream. Yet processed exports are, not surprisingly, less welcome in the consuming nations, which have traditionally carried out all manufacturing operations themselves. Trade policy conflicts seem certain to intensify rather than lessen in the years ahead.

The Middle Eastern countries' main bargaining strength in trade negotiations lies in the value of their markets to Western suppliers. For the European Community the Middle East represents a more important market than that of North America. The less populous O.P.E.C. countries of the Middle East all treat imports liberally, and have few quotas. Tariffs, where they exist, are modest, and there is complete freedom with regard to import payments. These countries have of course usually enjoyed balance of payments surplus, as the map indicates. In contrast, the more populous non-oil states, or the minor oil producers, such as the Sudan, Turkey or Egypt, habitually run balance of payments deficits. Shortages of foreign exchange have resulted in close control being exercised over all import payments. Nevertheless, despite such payments restrictions, countries attempting to industrialise such as Turkey or Egypt represent important markets for Western capital goods.

The future growth of the international trade of the Middle East will depend largely on how regional income rises, which in turn is affected by factors such as the future price of oil. The region's interdependence with the Western industrialised world implies, however, that trade growth will also be crucially influenced by the types of trading policies the Western countries themselves decide to adopt.

KEY REFERENCES

Sayigh, Yusif, *The Arab Economy: Past Performance and Future Prospects* (Oxford University Press, Oxford, 1982).

Wilson, Rodney, *Trade and Investment in the Middle East* (Macmillan, London, and Holmes and Meier, New York, 1977).

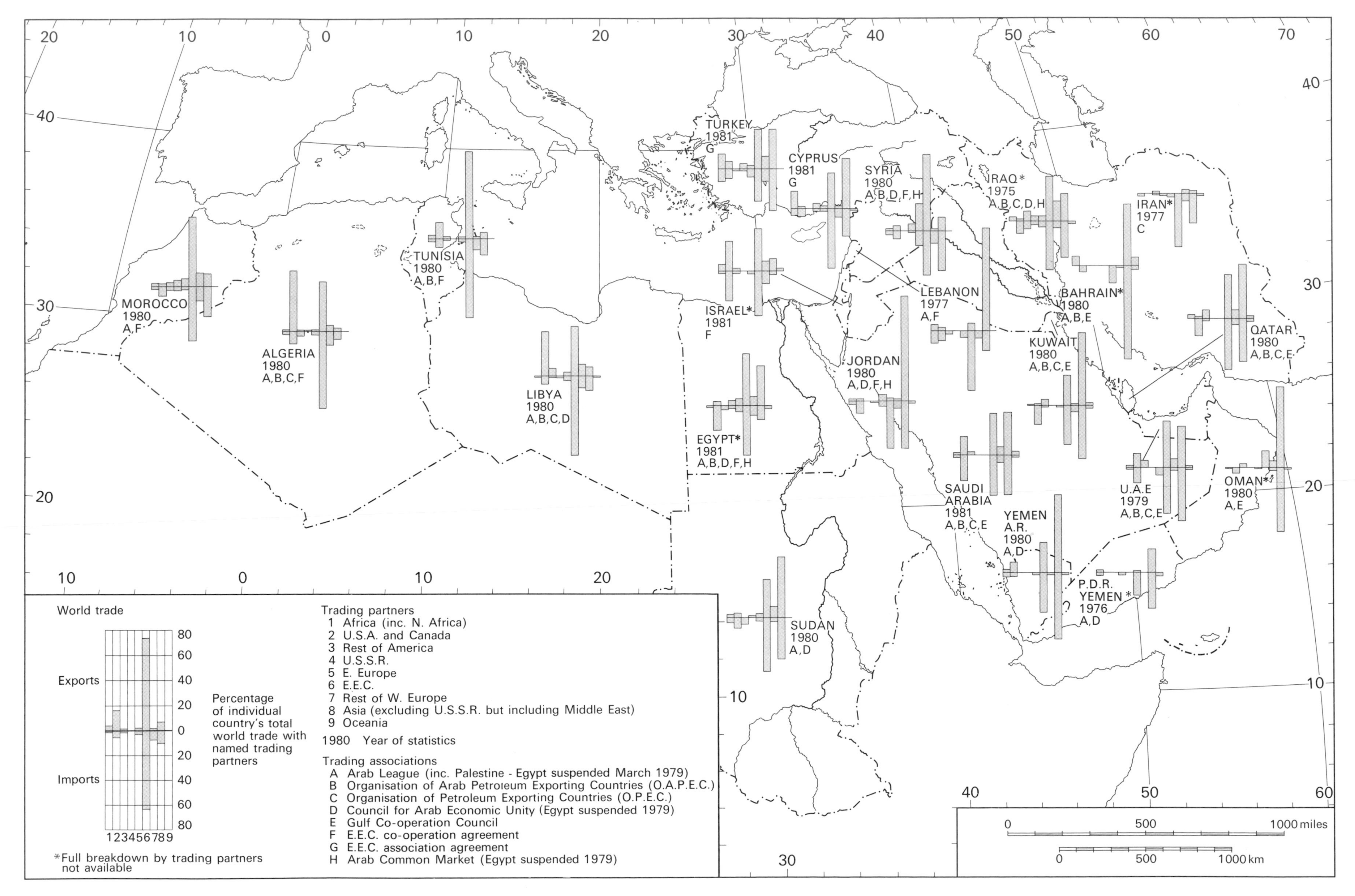

41 Trading partners and trading associations

Trading partners and trading associations

Rodney Wilson

As a general rule the direction of trade only alters over decades rather than years. The case of the Middle East seems to confirm this, despite the sweeping economic and political changes that have affected the region since World War II. In the last century, the region traded mainly with Western Europe, and this trading orientation continues despite attempts to promote greater economic linkages between the Middle Eastern countries themselves. The regional trading associations such as the Arab Common Market and the Gulf Co-operation Council have had little impact on the direction of trade, and it seems unlikely that they will unless conditions change. The Council for Arab Economic Unity established the Arab Common Market as part of a strategy to promote full Arab economic integration.

The Gulf Co-operation Agreement, of course, only dates from 1981 so it can be argued that it is premature to assess its impact. However, the Arab Common Market agreement was concluded in 1965, yet its impact on trade has been minimal. The motivation for joining was political rather than economic and it was seen as an extension of the Arab League, which encompassed all independent Arab states. The League is only a loose association of states, whereas a common market implies the full integration of its members' markets for traded goods. Such liberalisation means not only a close degree of economic co-ordination, but also some loss of economic and political sovereignty. Only Nasser's Egypt, Syria, Iraq and Jordan became full members of the Arab Common Market. Kuwait agreed to join, but later withdrew due to opposition from its parliament. Egypt was suspended in 1979 following its peace treaty with Israel.

Even apart from the political difficulties, the Arab Common Market had small chance of success. This was largely because the members had little to offer each other in trade. Egypt was the only state with an extensive manufacturing base, but its industries were geared to the domestic market, heavily protected, and internationally uncompetitive. Kuwaiti purchasers, faced with a choice between Egyptian-assembled cars and better quality, lower-priced Japanese models, naturally preferred the latter. Economic considerations were ultimately more important than benevolence towards fellow Arabs. Furthermore, the four states which joined the Arab Common Market all maintained payments restrictions because of shortages of foreign exchange. Although tariffs were abolished amongst member states in 1971, the payments controls remained, and in some cases became more restrictive.

The ability of the Gulf Co-operation Council to increase trade amongst its members must be considered limited. All the member states export oil and little else. Some petrochemicals may be marketed regionally, but most of the products from these new downstream industries are being sold in Japan and Western Europe. There are only likely to be modest petrochemical-consuming industries in the Gulf, despite the emphasis on the development of integrated industrial complexes. The only potential for intra-Arab trade in the Gulf is in supplies for the oil industry and construction materials. Most of the states have established a few consumer-oriented industries, but as they all manufactured the same kinds of products, such as soft drinks, it is difficult to see what advantage there would be in mutual exchange.

Middle Eastern trading relations with Western Europe are far from being satisfactory, despite the Euro-Arab dialogue and the co-operation agreements signed with the European Community. Oil imports enter the European Community market without restriction, but refined products and petrochemicals are subject to tariffs. The Community's variable import levies are applied to Middle Eastern agricultural exports in order to protect European farmers, while textiles are restricted by quotas under the multi-fibre arrangements. These restrictions also apply to Turkish clothing and footwear exports, even though the Ankara government signed an association agreement with the European Community providing for eventual accession. Basically, the greater the degree to which Middle East exports are processed, the tighter are the trade restrictions imposed by the European Community. This is frustrating for those who are striving to create employment in export-based industries in the Middle East and to increase local value added.

The European Community for its part alleges that, as the Middle Eastern states restrict trade themselves, they can scarcely argue for further trade concessions from Brussels. Most Middle Eastern governments have imposed tariffs to protect their infant industries and safeguard domestic employment. Furthermore the non-O.P.E.C. states all operate quota systems and foreign exchange licensing. Nevertheless, the Middle Eastern countries adhere to the principles of the General Agreement on Tariffs and Trade (G.A.T.T.) and do not discriminate against exports from particular countries, apart from the Arab trade embargo on Israel and, in the case of some states, Egypt.

As in other Third World markets, the European Community has been losing sales to Far Eastern suppliers. This is particularly the case in the eastern Arab states, where shipping goods from the Far East is in many cases cheaper than sending goods through the Suez Canal from Western Europe. Japanese cars dominate the markets of the Arabian peninsula, while civil engineering companies from South Korea and Taiwan have won major construction projects. Many capital and intermediate goods are supplied from the Far East, and Indian industries have also enjoyed notable successes in penetrating Middle Eastern markets. The trade with the Far East is two way, however, as these countries are major customers for oil and natural gas from the Gulf. Japan purchases most of the natural gas produced by the United Arab Emirates, and is the largest single customer for most of the Gulf O.P.E.C. states.

Despite superpower political involvement in the Middle East, Soviet and United States trade with the region is minimal outside the military field. The Soviet Union was a major supplier of both civilian and military equipment to Egypt in the 1960s, but its trade with most countries of the region, even Syria, has since declined. The United States mainly supplies armaments and aircraft to the region, but some of the trade is unrecorded for security reasons. Even where the items exported are known, as some are supplied as military aid on concessional terms it is difficult to estimate the value of such trade. The turbulent Middle East nevertheless represents the Third World's largest market for armaments, a facet of trading relations which many observers find disturbing.

KEY REFERENCES

Al-Mani, S. and Al-Shaikhly, S., *The Euro-Arab Dialogue* (Frances Pinter, London, 1983).

Ghantus, Elias T., *Arab Industrial Integration* (Croom Helm, London, 1982).

United Nations Economic Commission for Western Asia, *Economic Integration in Western Asia* (Frances Pinter, London, 1985).

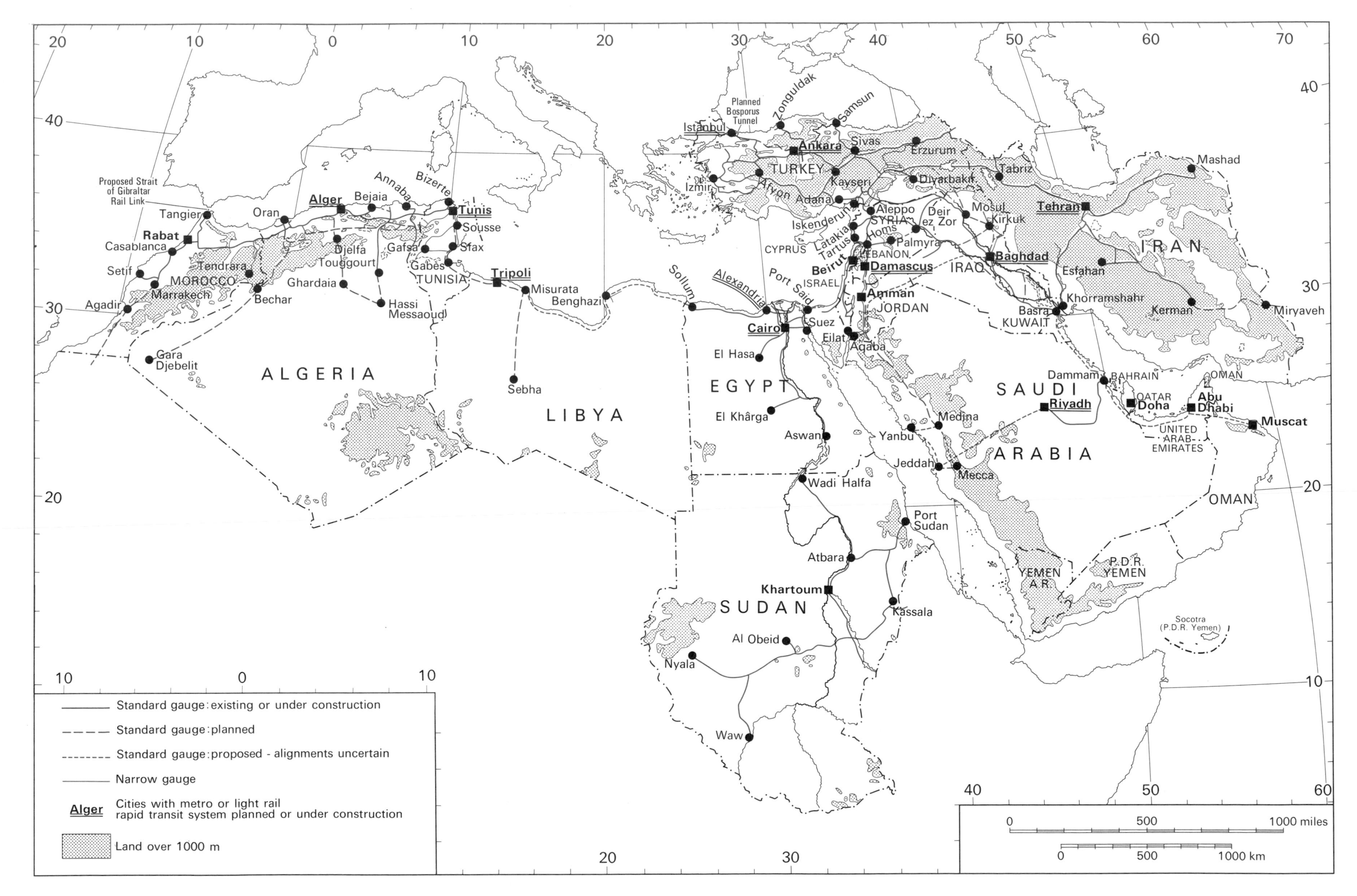

Proposed Strait of Gibraltar Rail Link
Planned Bosporus Tunnel
Tangier
Rabat
Casablanca
Setif
MOROCCO
Marrakech
Agadir
Tendrara
Bechar
Oran
Alger
Bejaia
Annaba
Bizerte
Tunis
Sousse
Sfax
Gafsa
Gabes
TUNISIA
Djelfa
Touggourt
Ghardaia
Hassi Messaoud
Gara Djebelit
ALGERIA
Tripoli
Misurata
Benghazi
Sebha
LIBYA
Sollum
Alexandria
Port Said
Cairo
Suez
El Hasa
EGYPT
El Khârga
Aswan
Wadi Halfa
Port Sudan
Atbara
Khartoum
SUDAN
Kassala
Al Obeid
Nyala
Waw
Istanbul
Zonguldak
Samsun
Ankara
Sivas
Erzurum
TURKEY
Izmir
Afyon
Kayseri
Adana
Diyarbakir
Iskenderun
Aleppo
SYRIA
Latakia
Tartus
Homs
Palmyra
Deir ez Zor
CYPRUS
LEBANON
Beirut
Damascus
ISRAEL
Amman
JORDAN
Eilat
Aqaba
Mosul
Kirkuk
IRAQ
Baghdad
Tabriz
Tehran
Mashad
IRAN
Esfahan
Kerman
Khorramshahr
Basra
KUWAIT
Miryaveh
Dammam
BAHRAIN
QATAR
Doha
Abu Dhabi
UNITED ARAB EMIRATES
OMAN
Muscat
SAUDI ARABIA
Riyadh
Medina
Yanbu
Jeddah
Mecca
YEMEN A.R.
P.D.R. YEMEN
Socotra (P.D.R. Yemen)
Standard gauge: existing or under construction
Standard gauge: planned
Standard gauge: proposed - alignments uncertain
Narrow gauge
Alger Cities with metro or light rail rapid transit system planned or under construction
Land over 1000 m
0 500 1000 miles
0 500 1000 km

42 Railways

Railways

Jonathan Mitchell

The current expansion and development of railways in the Middle East and North Africa is on a scale unmatched since the construction of the original rail networks in the early years of the twentieth century.

The early networks were far from comprehensive. The French left behind them the most complete systems, primarily in the Maghreb, but these have created problems for national rail companies by being an inconvenient mixture of gauges. With the exception of Iran and Iraq, the oil-rich states of today had little or no legacy from the earlier period of rail building and it is in these states that some of the most dramatic developments are taking place.

The increasing levels of investment reflect a view held throughout most of the region, that railways are vital to economic consolidation. Egypt, for example, increased its investment in railways tenfold in the 1970s and Turkey plans to invest some $1889 m over a ten-year period.

Development is occurring at two scales. The first is the creation of light rail rapid transit systems and metros for many of the region's increasingly overcrowded major cities. The second is at the scale of national networks, and the associated formation of international networks. Paramount at this scale is the construction of new track. In North Africa 1544 km of new track is currently under construction, with a further 10,400 km planned or proposed. In the rest of the region the figures are 1254 km and 9400 km respectively. In addition to this, replacement of old, frequently non-standard gauge track is a priority in a number of states, important to the cost efficiency of national networks. Rolling stock replacement and modernisation, traffic control system upgrading, and electrification are also major areas of investment.

The interlinking of national networks within the region and to those contiguous with it is a potentially important development. Current proposals allow for a link across the whole of North Africa, and, if the Gibraltar Strait tunnel is built, to Europe. The Bosporus tunnel will provide a valuable link to Europe for the north and east of the region, especially if proposals to link the Turkish and Iraqi rail systems come to fruition. Ultimately the whole of the Mashreq will be joined to both the Turkish system and the Arabian peninsula systems.

To ease the movement of international rail traffic and to co-ordinate development, two international bodies have been established in the region. In 1979 the Arab Union of Railways was established, and in 1984 Turkey, Syria, Iran, Iraq and Saudi Arabia joined to form the Middle East group of the International Union of Railways. In addition Turkey is already a member of the European container movement organisation.

Table 16 and Map 42 are designed to give a broad overview of the major rail developments in the Middle East and North Africa.

Table 16. *Railway developments*

	Development	Status
Algeria	64 km Alger light rail rapid transit	Advanced planning
	1000 km existing track renewal	Work progressing
	Gara Djebelit mineral line	Proposed
	High Plateau line	Advanced planning
Egypt	45 km Cairo regional metro	Operational 1985–6
	22 km Cairo Nile west bank suburban metro	Planned
	Alexandria metro	Feasibility study
	Cairo–Alexandria high speed train	Operational 1983
Iraq	32 km Baghdad metro	Planned
	100 km Baghdad–Husaibah link	Operational 1985
	282 km new track	Under construction
	2109 km new track	Planned
Iran	64 km Tehran metro	Work suspended
	Bandar Abbas–Bafq line	Planned
Jordan	Mineral line to Aqaba	Operational
	Link to Iraqi network	Planned
	Hejaz line reconstruction	Proposed
Libya	70 km Tripoli metro	Preliminary design
	170 km Tunisian border link	Planned
	2414 km new track	Planned/proposed
Morocco	160 k.p.h. electric link Rabat–Casablanca	Operational
	972 km Unity Line Marrakech–Layoun	Planned
	Rail tunnel to Spain	Under discussion
Saudi Arabia/ Arabian peninsula	Riyadh metro	Proposed
	Upgrading Riyadh–Dammam link	Operational 1984–5
	1200 km Riyadh–Jeddah–Mecca–Medina link	Operational 1987–8
	Hejaz line reconstruction	Proposed
	1600 km Basra–Muscat link	Proposed
Syria	Damascus metro	Preliminary study
	100% increase in network size	Operational 1984
	341 km existing track renewal	Planned
	Deir ez Zor–Iraq link	Operational 1987
	Latakia–Tartus link	Planned
	Tudmor–Deir ez Zor link	Proposed
Tunisia	30 km Tunis light rail rapid transit	Operational 1984–9
	1500 km existing track renewal	Planned
	Link to Libya	Planned
Turkey	26 km Ankara light rail rapid transit	Planned
	12+ km Istanbul metro	Planned
	319 km existing track renewal	Work progressing
	416 km Istanbul–Ankara new link	One-third complete – delayed
	Link to Iraq	Under discussion
	9 km Bosporus tunnel	Proposed
	Local locomotive assembly	Current

Sources: Briginshaw, D., *et al.*, 'Railways and railway engineering: business feature', *M.E.E.D.*, 3 Aug. 1984, pp. 25–36. Various articles in *M.E.E.D.*, *Middle East Construction*, and *Middle East*, 1983–5.

KEY REFERENCES

Briginshaw, D. *et al.*, 'Railways and railway engineering', business feature, *M.E.E.D.*, 3 Aug. 1984, pp. 25–36.

Various articles in *M.E.E.D.*, *Middle East Construction*, and *Middle East*, 1983–5.

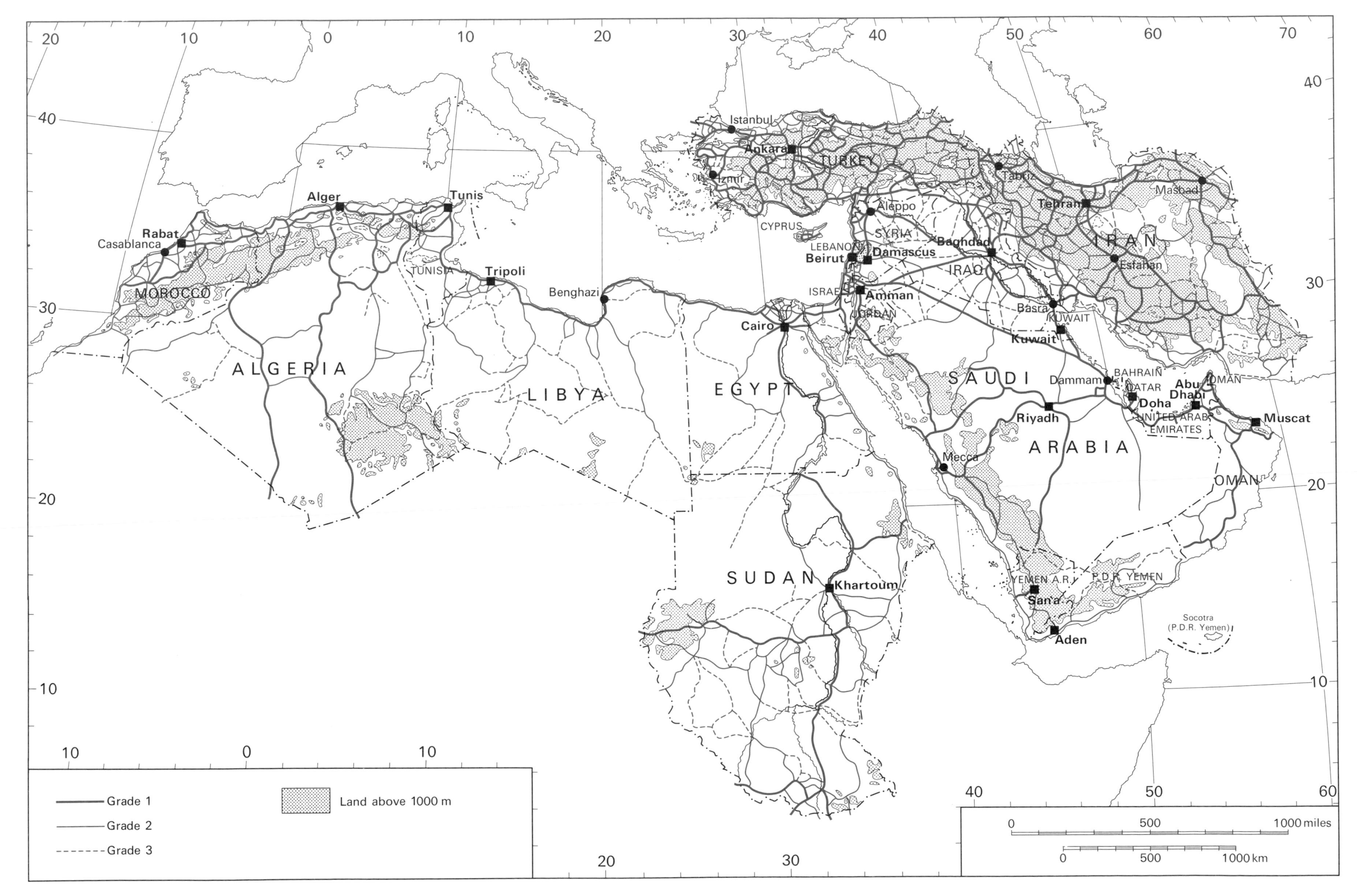

Istanbul
Ankara
TURKEY
Izmir
Tabriz
Mashad
Tehran
IRAN
Esfahan
Aleppo
CYPRUS
SYRIA
LEBANON
Beirut
Damascus
Baghdad
IRAQ
ISRAEL
Amman
JORDAN
Basra
KUWAIT
Kuwait
Alger
Tunis
Rabat
Casablanca
TUNISIA
Tripoli
Benghazi
MOROCCO
ALGERIA
LIBYA
EGYPT
Cairo
SAUDI
ARABIA
Dammam
BAHRAIN
QATAR
Doha
Abu Dhabi
OMAN
UNITED ARAB EMIRATES
Muscat
Riyadh
Mecca
SUDAN
Khartoum
YEMEN A.R.
San'a
P.D.R. YEMEN
Aden
Socotra (P.D.R. Yemen)
Grade 1
Grade 2
Grade 3
Land above 1000 m
0
500
1000 miles
0
500
1000 km

43 Roads

Roads

Jonathan Mitchell

For almost 3000 years, the Middle East has been one of the world's great crossroads. Across it generations of merchants and pilgrims, and the ebb and flow of empires, have left a network of roads and routeways. The Romans were the first to establish a region-wide network of paved roads – the key to the control of their sprawling empire. Later, the Ottoman Empire expanded the system as did the twentieth-century colonial powers. But what they left behind them was generally inadequate for the pace and type of recent road transport developments.

Since World War II the region has experienced a motor vehicle explosion, both private and commercial. Even Egypt, one of the poorer countries of the region, currently has a 12% per annum increase in road traffic. To serve the increase in road traffic, and to service the economic development, which has itself in part stimulated the explosion, the old road networks are being upgraded and considerably expanded. The most spectacular developments at all levels, from urban road systems to international expressways, are taking place in the hydrocarbon-rich states, but it should be noted that it was these states which had the least comprehensive networks to begin with.

Urban roads have undergone and are undergoing something of a revolution. Few, if any, of the old towns and cities of the region have roads suited to motor traffic, let alone the high volumes of traffic to which they are now subjected. Some cities have experienced considerable changes for the benefit of motor vehicles. Reza Shah bulldozed new boulevards through the heart of old Shiraz, and a similar fate befell parts of Istanbul, though originally in the interests of civil order.

Newer parts of many cities, built with the motor vehicle in mind, are now frequently badly congested. This has provoked two key responses. The first has been to plan or build light rail rapid transit systems and metros (see Map 42). The second has been to build ring roads and bypasses.

Egypt's overcrowded cities are prime examples of the need for ring roads. The problem, apart from the general paucity of funds, is the undesirability of encroachment on valuable agricultural land. For this reason the delta towns are unlikely to get ring roads, but congestion is now so acute in Cairo that it cannot do without one. Japan is currently financing a feasibility study for such a road, which will run, at least in part, on the desert fringe of the city. This should also serve to aid the establishment of overspill settlements in the desert, away from the agricultural land.

In the Arabian peninsula states the problem of land use clash does not apply, and neither does the problem of complex land ownership which plagues road development in the older towns and cities of the region. Kuwait has so far built six ring roads as the city has grown. Saudi Arabia has given top priority to ring road construction, six projects being in hand. The largest of these is the 90 km Riyadh ring road. Abu Dhabi has recently completed a multi-million dollar 1 km underpass which, linked into the existing urban and suburban road system, gives the city, it is claimed, the best road network in the Middle East.

It is, however, at the national network and international network levels that the most dramatic developments are taking place. In 1954 Saudi Arabia had just 237 km of paved road. By the end of 1985 this figure had risen to around 27,000 km. Of this, 4000 km is expressway. Next to Saudi Arabia, Iraq has the most ambitious road-building programme: 1830 km of expressway is under development, and should link the country into the planned Turkish motorway system, the Kuwaiti and Arabian peninsula systems, and the road networks of Syria and Jordan.

The Turkish road development programme is experiencing funding difficulties, and only if her Arab neighbours provide the means is the planned motorway system likely to achieve fruition.

In North Africa, Libya alone has a major road-building programme. A coastal expressway is under construction, as are a number of new roads into the interior. Sudan and Egypt are both short of money, although Sudan has been able to increase its metalled road length more than fourfold since the end of the 1970s. In Egypt the upgrading of major trunk routes is seen as vital to overall development. The Cairo–Alexandria desert highway is being upgraded to six lanes, whilst the Alexandria to Mersa–Matruh road is being upgraded to four lanes. In addition, important new link roads are at last being constructed outside the Nile valley.

Road-building programmes in the Maghreb are being given less importance than elsewhere. Both Morocco and Algeria are emphasising rail transport on the one hand, and on the other their colonial road network legacy is far more comprehensive than elsewhere. However, the Trans-Saharan highway in Algeria is a considerable new development.

Ultimately the whole of the region should be linked by good quality highways. The North African states are already linked as a result of the colonial period, and it should not be long before the proposed new Egypt–Saudi Arabia and Mashreq link is established. Saudi Arabia and Bahrain are now joined by a 25 km causeway, a 90-minute drive on motorway-standard road. Proposals have been put forward for another causeway, this time 70 km long to join Bahrain to Qatar.

A number of projects linking the region to Europe and Africa are under way. The first Bosporus Bridge is to be joined by a second in the 1990s, greatly easing access to and from Europe for the north of the region. In the west of the region a fixed link across the straits of Gibraltar is under discussion. In 1982 Spain and Morocco agreed that this should be a rail tunnel, but presumably this would cater for road traffic in a way similar to the proposed English Channel Tunnel. To the south the Trans-Saharan highway will link Algeria to Niger, and Sudan is building a 530 km road to Kenya.

A feature of international developments has been a lack of co-ordination between national road-building programmes. To help combat this problem the Arab Overland Transport Corporation was set up in 1981 by the Council for Arab Economic Unity, but its effects are as yet unclear.

Overall, the development of roads within the Middle East and North Africa is on a scale not matched anywhere else in the developing world. This is a situation which is unlikely to change for some considerable time. A comprehensive road network throughout the region is vital to its long-term economic growth.

KEY REFERENCES

Roberts, J., 'Road projects in the Middle East', *Middle East Review*, 1983, pp. 69–72.

Various articles from *Middle East Construction*, 1985.

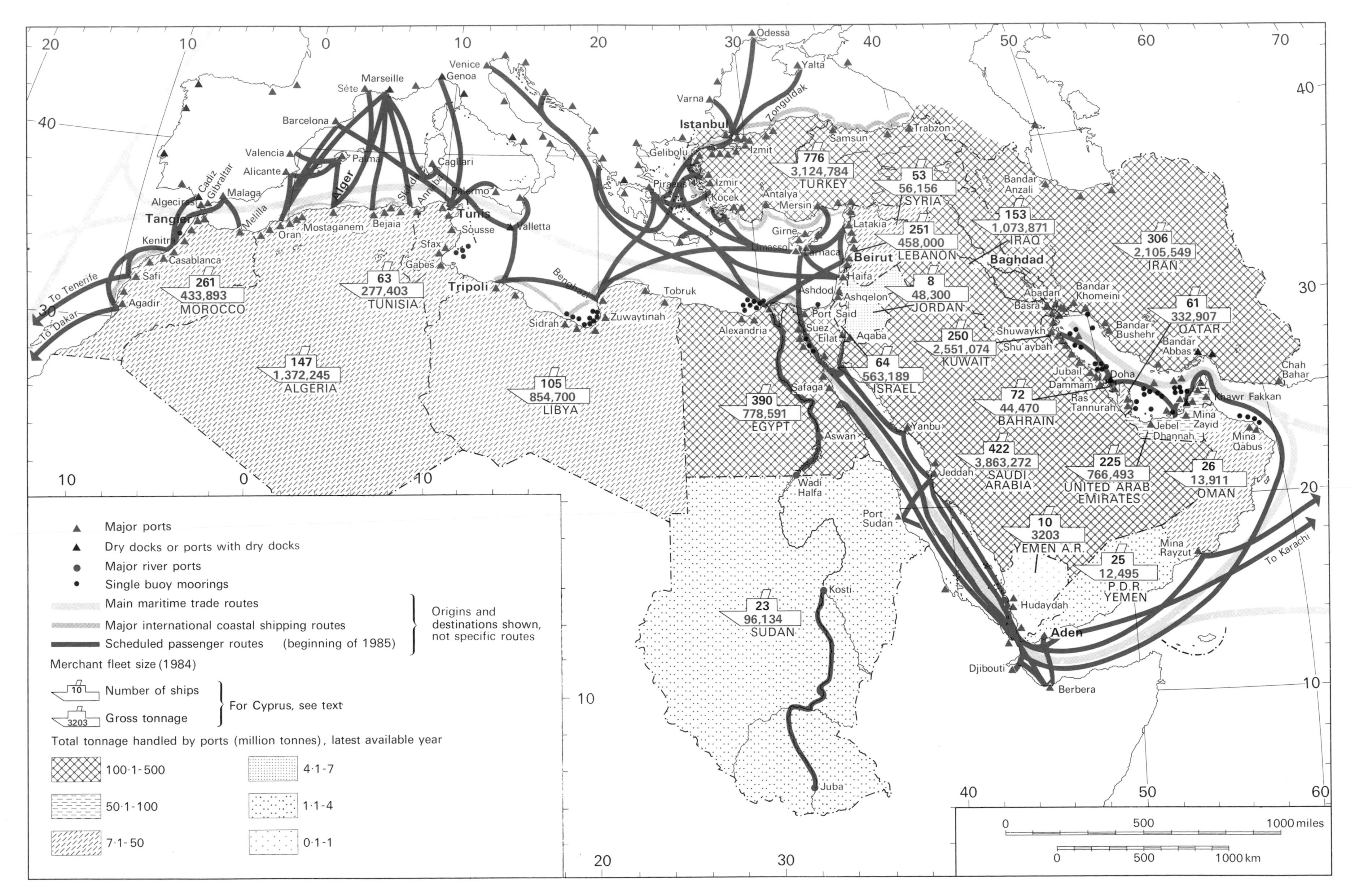

44 Ports and shipping

Ports and shipping

Gerald Blake

The Middle East and North Africa are favourably located to engage in regional and international maritime trade. The Mediterranean region, the Gulf and parts of Arabia enjoyed considerable prosperity centuries ago because of their trading activities by sea. With the opening of the Suez Canal in 1869 (Map 50) the region straddled one of the prime global shipping routes, which acquired increasing strategic and economic significance, particularly after World War II, because of its role in the movement of oil (Map 46). Coastal shipping is also important, particularly where internal distances within states are considerable and communications difficult. The map is too small to indicate the complexity and intensity of coastal trade involving small vessels, such as the picturesque *dhows* of the Gulf and Indian Ocean, but certain scheduled coastal services are illustrated for Turkey, Algeria and Morocco. Navigable rivers are relatively unimportant, though river steamers and barges are much used on the Nile to Wadi Halfa, and between Kosti and Juba in Sudan. The lower Shatt al Arab river is navigable by large ships as far as Basra, thus providing Iraq with an invaluable outlet before the Gulf war began in 1979.

There are approximately 140 major ports in the region in addition to oil-exporting terminals and offshore mooring buoys for tanker loading. The pattern of port location reflects population distribution and economic activity. Port locations have changed considerably over the centuries as political and economic fortunes have fluctuated. Some formerly rich and famous ports such as Ephesus in Turkey are today silted up and in ruins. The creation of new political units after World War I and the political fragmentation of the Gulf region have undoubtedly stimulated the growth of more ports than are strictly necessary for the needs of the region. The impact of political events upon port activity is well illustrated by the Iran–Iraq war, which has effectively knocked out Iraqi outlets to the Gulf, channelling much of Iraq's external trade through the Jordanian port of Aqaba. Similarly, Iran is planning to expand ports at Bandar Abbas and Chah Bahar in the lower Gulf, away from the war zone.

Port construction, expansion and modernisation have been carried out in many parts of the Middle East and North Africa since the early 1970s on a scale probably unequalled anywhere in the world. This was partly in response to a period of massive growth in the volume of imports, particularly construction materials such as cement, which had created serious congestion and unloading delays of from 50 to 150 days. The worst affected ports were around Arabia, but long delays were typical throughout the region, including notably Libya and Iran. Several colossal projects were undertaken, including the construction of whole new industrial and port complexes near Yanbu and at Jubail in Saudi Arabia. The most extraordinary developments have been in the United Arab Emirates, where each federal state has caught port-building fever, resulting in considerable over-capacity and a total of 134 deepwater berths by 1985. One notable feature of port modernisation as a whole is the widespread introduction of specialist facilities such as container ports, roll-on/roll-off berths, bulk handling facilities, and livestock terminals as at Jeddah and Yanbu. But not all the region's ports are efficient or modern, as the relief agencies discovered when bringing supplies to Sudan in 1985. Nor are all the port-building projects yet complete. New ports were still being constructed at Jilil (Algeria), Damietta (Egypt), and Nishtun (P.D.R. Yemen) in 1986. There seems little doubt that competition between ports will intensify in future for the servicing of ships in transit, handling transit goods, and normal import–export activities.

Although the region has an ancient sea-going tradition, domestic merchant fleets remained poorly developed until recent decades, especially in the Arab world. Even in 1985 the combined merchant fleets of seven states (Oman, Jordan, Syria, Sudan, Bahrain and the Yemens) total only about 285,000 gross tonnage, and together these states possess only some 220 ships of over 100 gross tonnage. Since the early 1970s, however, several states have made strenuous efforts to expand their merchant shipping interests, in the first instance, as a way of obtaining a greater share in the downstream marketing of crude oil and petroleum products. Thus the oil-rich states, with surplus revenues to spend, led the way. A number of complex joint ventures were embarked upon, including the Arab Maritime Petroleum Company (A.M.P.T.C.), founded in 1973 by Bahrain, Saudi Arabia, Kuwait, Algeria, Qatar, Iraq, Libya and Abu Dhabi, which was an O.A.P.E.C. initiative to encourage Arab oil transportation. A.M.P.T.C. remains one of the largest shipping companies in the Middle East. A second venture was started in 1976 with the formation of the United Arab Shipping Company (U.A.S.C.) by Kuwait, Saudi Arabia, the United Arab Emirates, Bahrain, Qatar and Iraq, designed to carry dry cargoes as well as oil. The U.A.S.C. is still a powerful force in shipping, second only in the region to the state-owned Kuwait Oil Tanker Company (K.O.T.C.), a subsidiary of the Kuwait Petroleum Corporation. Ironically, much of this expansion began as world shipping entered a period of recession, and orders for new ships from the Middle East were good news for flagging shipyards in Europe and elsewhere. The merchant fleets of the region showed net expansion until about 1983 when the effects of world recession began to have serious consequences. Long before 1983, however, some companies were in great difficulty, particularly those heavily dependent upon the oil industry.

Table 17. *Leading oil tanker fleets*

	Ships	Gross tonnage (1984)
Cyprus	69	3,180,146
Saudi Arabia	96	1,950,528
Kuwait	26	1,430,825
Turkey	77	1,165,415
Iran	31	917,680
Iraq	28	797,055
Libya	15	745,056
Algeria	22	593,987
United Arab Emirates	24	513,417

Source: Lloyds of London, *Lloyds Register of Shipping, Statistical Tables* (Lloyds, London, 1984).

In 1984 the merchant fleets of the Middle East and North Africa, excluding Cyprus, represented 2.7% of the world total tonnage, compared with 2.5% a decade before. Cyprus is used as a flag of convenience, and the number of ships and tonnage are by far the largest in the region (737 ships totalling 6,727,887 gross tonnage). After Cyprus, the largest fleets by tonnage are those of Saudi Arabia, Turkey, Kuwait, Iran and Algeria, and by numbers of ships the largest fleets are those of Turkey, Saudi Arabia, Egypt and Iran (see map). All the major merchant shipping states of the region are heavily involved in oil transportation, and the region as a whole controls nearly 8% of the world's tanker fleet (Table 17), compared with 2.2% a decade earlier. The prevalence of oil tanker traffic has underpinned the growth of ship repairing activities and dry docks in the region. Thus O.A.P.E.C. sponsored the construction of one of the world's largest dry docks in Bahrain in 1977, capable of accommodating supertankers of approximately 300,000 gross tonnage. In recent years the Bahrain shipyard has had difficulty attracting sufficient trade in spite of its strategic location. Kuwait, Dammam and Jeddah (in Saudi Arabia) and Port Said also offer repair and small-scale shipbuilding facilities.

KEY REFERENCES

Arab Shipping Guide (Seatrade Publications, London, 1985).
Couper, A. D. (ed.), *The Times Atlas of the Oceans* (Times Books, London, 1983).
Lloyds Ports of the World 1985 (Lloyds of London Press, London, 1985).

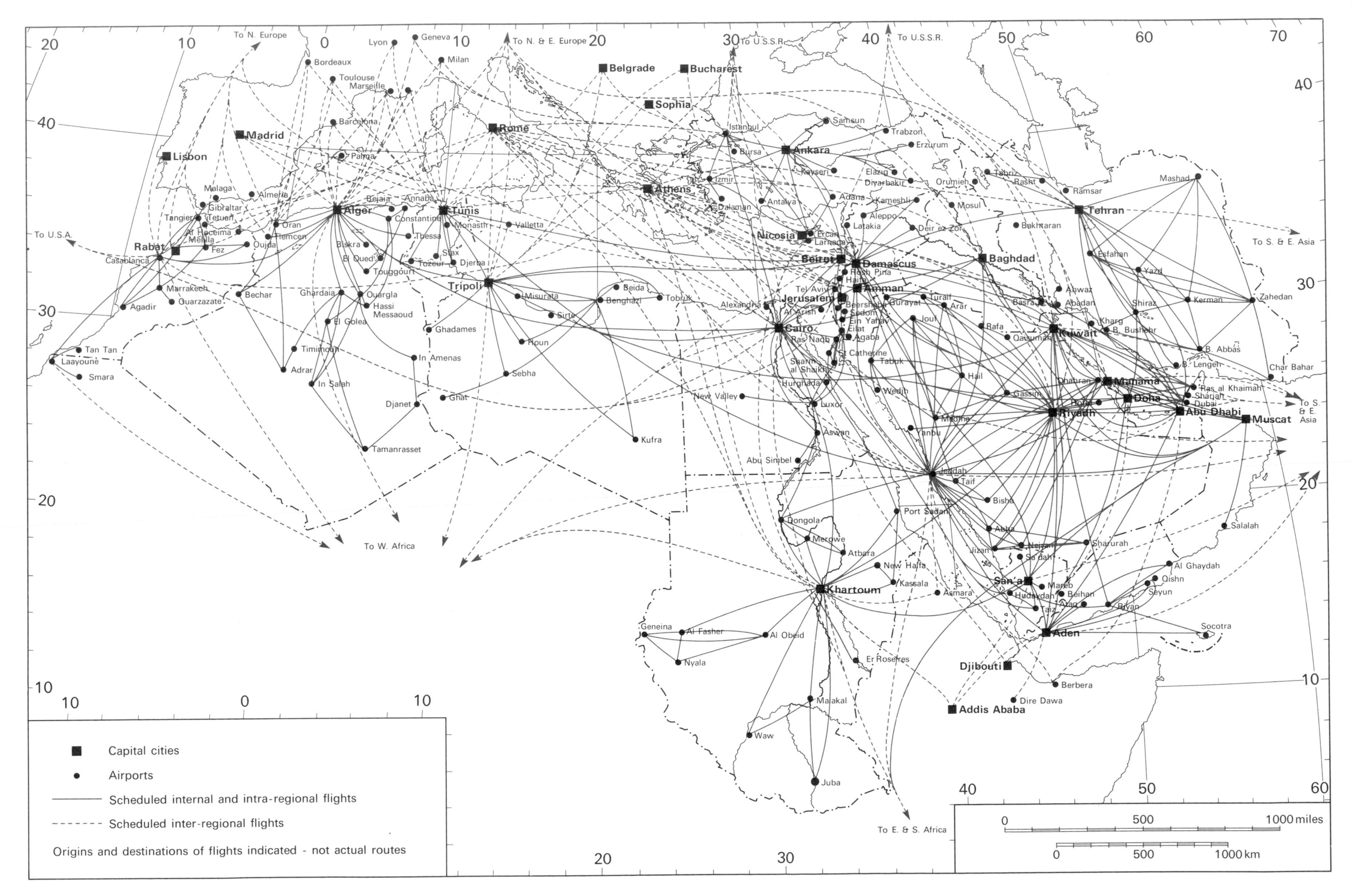

To N. Europe
Lyon
Geneva
To N. & E. Europe
To U.S.S.R.
Bordeaux
Toulouse
Marseille
Milan
Belgrade
Bucharest
Sophia
Barcelona
Rome
Madrid
Lisbon
Palma
Istanbul
Samsun
Trabzon
Erzurum
Ankara
Bursa
Kayseri
Elazig
Diyarbakir
Izmir
Athens
Malaga
Almeria
Gibraltar
Tangier
Tetuan
Al Hoceima
Melilla
Fez
Oujda
Oran
Tlemcen
Bejaia
Annaba
Constantine
Alger
Tunis
Monastir
Valletta
Tebessa
Sfax
Tozeur
Djerba
Biskra
El Oued
Touggourt
To U.S.A.
Rabat
Casablanca
Marrakech
Ouarzazate
Agadir
Bechar
Ghardaia
Ouargla
Hassi Messaoud
El Golea
Timimoun
Adrar
In Salah
In Amenas
Ghadames
Djanet
Ghat
Tamanrasset
Tan Tan
Laayoune
Smara
Tripoli
Misurata
Sirte
Benghazi
Beida
Tobruk
Houn
Sebha
Kufra
To W. Africa
Dalaman
Antalya
Adana
Kameshli
Aleppo
Latakia
Deir ez Zor
Mosul
Orumieh
Tabriz
Rasht
Ramsar
Mashad
Tehran
Bakhtaran
Nicosia
Ercan
Larnaca
Beirut
Damascus
Rosh Pina
Haifa
Tel Aviv
Jerusalem
Amman
Alexandria
Al Arish
Beersheba
Sodom
Ein Yahav
Eilat
Aqaba
Cairo
Ras Naqb
St Catherine
Sharm al Shaikh
Hurghada
Luxor
Aswan
Abu Simbel
New Valley
Turaif
Arar
Gurayat
Jouf
Tabuk
Wedjh
Madina
Yanbu
Hail
Gassim
Rafa
Qasumah
Baghdad
Basra
Ahwaz
Abadan
Kuwait
Kharg
B. Bushehr
Esfahan
Yazd
Shiraz
Kerman
Zahedan
B. Abbas
B. Lengeh
Char Bahar
Dhahran
Manama
Doha
Ras al Khaimah
Sharjah
Dubai
Abu Dhabi
Muscat
Riyadh
To S. & E. Asia
Jeddah
Taif
Bisha
Abha
Nejran
Jizan
Sa'dah
Sharurah
Salalah
Port Sudan
Dongola
Merowe
Atbara
New Halfa
Kassala
Khartoum
Asmara
San'a
Mareb
Hudaydah
Beihan
Ataq
Taiz
Riyan
Al Ghaydah
Qishn
Seyun
Socotra
Aden
Djibouti
Berbera
Dire Dawa
Addis Ababa
Geneina
Al Fasher
Al Obeid
Nyala
Er Roseires
Malakal
Waw
Juba
To E. & S. Africa
Capital cities
Airports
Scheduled internal and intra-regional flights
Scheduled inter-regional flights
Origins and destinations of flights indicated - not actual routes
0
500
1000 miles
0
500
1000 km

45 Airways

Airways

Jonathan Mitchell

Since the end of World War II a comprehensive civil aviation system of air routes, airports, and airlines has developed in the Middle East and North Africa, providing both intra- and inter-regional air links. The current level of development is the product of three key factors.

First, the position of the region as a major international crossroads means that many flights transit the region, generally landing at least once within it. A number of airports, most notably in the Gulf, have been built essentially to attract this through traffic. One result in the Gulf is an extraordinary number of modern international airports in a very small area.

Secondly, a very important factor is the position of the region in world trade. The oil-rich states, especially, are major trading partners with the Western nations. Indeed, aircraft themselves are major trade items. In 1984 Saudia purchased ten Boeing 747-300s at a cost of $1000 million. Sales of the A300-310 Airbus are worth around $2000 million to Europe. The number of routes and flights to these states from Western Europe, North America and Japan reflect the business traffic generated.

Thirdly, at the level of intra-regional and domestic air services, a highly significant factor is the sheer size of many of the countries. The distances between centres are often considerable and land transport, though improving, is frequently slow. An indication of the importance of domestic civil aviation can be gained from the figures for the average growth of domestic passenger traffic between 1972 and 1984: 22.6% in the Middle East and North Africa, against a world rate of 6.2% For the period 1982–92 a figure of 11% is predicted for the region, while 6% is predicted for the rest of the world.

The Middle East and North Africa is served by a great many international airlines. Most of the world's airlines have scheduled flights to and from the region. In addition, all the 22 states in the region have a national airline (Table 18). National airlines, whether privately or state-owned, are considered to be prestige organisations, but few have international reputations for excellence. Of those shown in Table 18 Gulf Air has perhaps the best reputation amongst businessmen and other travellers alike. A number run at a loss but are kept going by injections of state capital, simply because they are important to the self-image of many states, particularly where international air routes are involved. The number of regional carriers has produced considerable duplication of services and this is one of several fields for inter-state co-operation under discussion. The fact that one airline can serve the needs of more than one state is clearly shown in the case of Gulf Air (Table 18) which serves four states.

Further areas for co-operation suggested by the Arab Air Carriers Organisation are that each national carrier should specialise in the maintenance of a particular aircraft type, and that airlines should work together when new equipment is being purchased. Also, standardisation of air traffic control has been suggested. There are many advantages to be gained from co-operation, cost reduction overall and a reduced reliance on expatriate pilots and ground crews being two of the most important.

Throughout the region there has been considerable investment in airport construction, both major international airports and domestic airports. Efficiency, security and appearance of the former are given much emphasis. The region now possesses some of the most up-to-date airports in the world. Perhaps the most striking airport feature in the region is the new Haj terminal at Jeddah, built to be used specifically by pilgrims during the Haj. Elsewhere airport developments may be less striking, but they are of great importance.

The majority of the aircraft operated by Middle Eastern and North African airlines are of U.S. origin. The Boeing Company dominates the civil airliner market here as in much of the non-Communist world. However, Europe is making inroads in the airliner market, and a few states such as Libya, Syria and Iraq operate Soviet-built aircraft in addition to American and European types.

Table 18. *Major Middle Eastern and North African airlines*

Country	Airline	Type of service*	Date of original formation
Algeria	Air Algérie	PCI	1947
	Inter Air Services	PCD	1984
Bahrain, Qatar, Oman, U.A.E.	Gulf Air	PCI	1974
Cyprus	Cyprus Airways	PCI	1947
Egypt	Egypt Air	PCI	1932
	Misrair	PCD	1964
Iran	Iran Air	PCID	1944
Iraq	Iraqi Airways	PCID	1945
Israel	El Al	PCI	1948
Jordan	Alia	PCID	1963
Kuwait	Kuwait Airways	PCI	1953
Lebanon	Middle East Airways (MEA)	PCI	1945
	Trans-Mediterranean Airways	CI	1953
Libya	Libyan Arab Airlines	PCID	1964
Morocco	Royal Air Maroc	PCID	1946
Oman	Oman Aviation Services	PCID	1981
Saudi Arabia	Saudia	PCID	1945
Sudan	Sudan Airways	PCID	1945
Syria	Syrian Air	PCID	1946
Tunisia	Tunis Air	PCID	1948
Turkey	Türk Hava Yolları (THY)	PCID	1933
Yemen A.R.	Yemenia	PCID	1951
Yemen, P.D.R.	Alyemda	PCID	1971

*P = Passenger; C = Cargo; I = International; D = Domestic
Source: D. Donald (ed.), *The Pocket Guide to Airline Markings and Commercial Aircraft* (Temple Press/Aerospace, Feltham, 1985).

Air routes matrix

The cartographic representation of the number of air routes converging on the airports within the Middle East and North Africa (Map 45) gives a reasonable, if one-sided, picture of the relative importance of individual airports. To complete the picture, it is necessary to refer to the left-hand side of the matrix (Fig. 5), and Table 19. The upper part of the matrix shows the number of scheduled flights using airports within the region, per week, at the end of 1985. The lower portion of the matrix shows the number of scheduled flights operating between the Middle East and North Africa and selected airports outside the region. Table 19 shows the total number of flights between the airports included in the matrix.

For the most part, the oil-rich states have better-developed air linkages than those which are poorer and oil deficient. Saudi Arabia and P.D.R. Yemen are the two extremes. The distances involved for travel between and around the oil-rich states (see the right-hand portion of the matrix), and the relative ease and speed of travel by air as opposed to road or rail, are perhaps as significant as the prosperity of the states. Iran, however, despite its oil, is the worst-linked state, and this clearly reflects its political isolation.

Certain airports stand out as being particularly important in terms of the number of scheduled flights handled. Jeddah and Cairo rank first and second, both for the total and for intra-regional flights handled. Cairo is also the best-linked airport to destinations outside the region, but Jeddah drops to ninth place in this category. The explanation shows clearly in the matrix. Ninety of the flights in and out of Jeddah link it to Riyadh, reflecting the *de facto* twin capital status of the cities within Saudi Arabia. The importance of Cairo clearly arises from Egypt being the most populous state in the region, one of the most industrially developed states, and from its status as a major tourist destination. Tel Aviv features in second place in the inter-regional category, but ranks nineteenth within the region,

Distances between airports in statute miles† (above diagonal); Average number of flights between airports per week* (non-transfer) (below diagonal)

	Abu Dhabi	Aden	Alger	Amman	Baghdad	Beirut	Cairo	Casablanca	Damascus	Doha	Istanbul	Jeddah	Khartoum	Kuwait	Manama	Riyadh	Tel Aviv	Tehran	Tripoli	Tunis	Athens	Lagos	London	Moscow	Nairobi	New York	Paris	Rome	Singapore	Sydney	Tokyo
Abu Dhabi		1012		1232		1314	1468	3755	1248	191	1867	992	1546	520	271	489			2555	2741	2029		3413	2322	2133		3260	2679	3664	7527	5027
Aden	2					1576	1480		1521	960		712	858	1150		830								3020	1124		3475				
Alger	9						1681	192	1888			2395		2642		2691			630	389	1129		1025	2069			841	605			
Amman	18				492	149	295	2523	121	1044	754	734		734	961	811			1335	1509	816		2280	1659		5732	2104	1473	4878		
Baghdad				15		506	785	2972			1001	868	1421	355	617	608				1938	1194		2545	1582			2393	1829			5556
Beirut	4	3		3	0		351		66	1132	614	882	1358	802	1045	923				1435	716	2786	2154	1550		5609	1983	1369			
Cairo	9	1	4	21	27	5		2289	377	1277	764	769	1007	997	1200	1010	244			1302	694	2443	2185	1790	2204	5608	1995	1329	5130		6198
Casablanca	1		5	1	1		2		2527			2959		3257		3298			1201	1034	1778		1293	2636		3609	1188	1237			
Damascus	3	2	3	6		2	5	2		1066	672	840		736	979	860		857	1351	1500	782		2216	1538			2046	1435			
Doha	28	1		6		2	9		2			826	1399	352	91	307					1847		3248				3091				
Istanbul	1			4	8	1	3		6			1477		1349		1523	706	1270	1037		345		1554	1089		5009	1394	852			
Jeddah	5	7	8	8	3	7	37	6	8	4	6		597	766	794	526			1772	2032	1461	2637	2952		1584	6371	2763	2094	4564		5957
Khartoum	3	1			2	2	18			3		24		1361		1112				2309	1633		3062	2797	1207		2937	2290			
Kuwait	16	4	2	14	7	6	14	2	14	8	2	13	2		262	322			2069	2236	1514		2898	1937		6345	2742	2162	4145		5201
Manama	29			8	0	2	4		5	39		10		21		268				2470	1760		3158				3001	2412	3930	7798	5167
Riyadh	3	1	2	4	2	0	14	5	5	4	3	90	2	9	9						1627		3074			6530	2905	2284	4138		
Tel Aviv							10				2										752		2222		2306	5672	2044	1409			
Tehran									4		3										1530		2738	1526			2610	2121			5520
Tripoli	2		3	3				9	9		11	3		3						333	697	1911	1457	1985			1249	632			
Tunis	5		10	1	1	1	4	7	3			5	2	2	1				5		744		1126	1824			918	362			
Athens	3		1	7	3	5	10	2	6	1	12	5	4	4	4	4	8	1	10	2											
Lagos						1	2												2												
London	22		4	8	4	4	16	4	3	9	10	11	5	15	29	9	10	3	6	5											
Moscow	3	3	4	1	2	1	1	2	4		0	0	0	0				1	8	1											
Nairobi	1	1					3					4	5				1														
New York				5		0	12	2			8			6		6	23														
Paris	9	1	49	5	4	7	17	26	6	3	9	9	2	5	5	7	22	3	2	16											
Rome	0		15	4	3	2	19	5	5		10	5	1	4	4	4	12	2	12	14											
Singapore	7			2			2					4		2	15	2															
Sydney	1														8																
Tokyo	1				1		5					1		2	3			2													

*0 = Occasional non-transfer flights
†Direct route, or most direct route actually flown, to include stops

Fig. 5 Middle East and North Africa scheduled flight matrix. Where data is not shown, scheduled air routes are not in operation. Note that London = Heathrow and Gatwick. *Source: A.B.C. World Airways Guide* (A.B.C. Travel Guides, Dunstable, Dec. 1985).

Table 19. *Total number of scheduled flights per week between selected airports*

Airport	Inter-regional	Rank	Intra-regional	Rank	Total	Rank
	Number of scheduled flights					
Abu Dhabi	47	6	138	5	185	4
Aden	5	20	22	18	27	19
Alger	73	3	46	15	119	8
Amman	32	12	112	7	144	7
Baghdad	17	16	66	10	83	15
Beirut	20	15	38	17	58	18
Cairo	87	1	187	2	274	2
Casablanca	41	7	41	16	82	16
Damascis	24	14	79	9	103	10
Doha	13	18	106	8	119	8
Istanbul	49	5	50	12	99	11
Jeddah	39	9	244	1	283	1
Khartoum	17	16	59	11	76	17
Kuwait	38	10	139	4	177	6
Manama	68	4	128	6	196	3
Riyadh	32	12	153	3	185	4
Tel Aviv	76	2	12	19	88	12
Tehran	12	19	7	20	19	20
Tripoli	40	8	48	13	88	12
Tunis	38	10	47	14	85	14

Source: A.B.C. World Airways Guide (A.B.C. Travel Guides, Dunstable, Dec. 1985).

perhaps not surprisingly given the geopolitics of the Middle East and North Africa. Its only links within the region are to Cairo and Istanbul.

Manama is the third ranked airport overall. This is because Bahrain is both an island and a major business centre within the Gulf region. Much of the traffic is made up of short-distance flights from other Gulf states, but Bahrain has also been successful in winning much of the East–West through-flight refuelling business. It is interesting, however, to speculate whether the overall importance of Manama airport will decline with the opening of the road causeway to Saudi Arabia.

Two major cities outside the region are particularly well linked with it. These are London and Paris. The matrix shows that London is the main European centre for flights to and from the east of the region, whilst Paris is the main centre for flights to and from North Africa. The degree of air linkage between Paris and Alger is noteworthy – 49 scheduled flights per week. This must reflect the close relationship between France and her former colony, Algeria.

KEY REFERENCES

Coombs, T. and Williamson, J., 'Middle East aviation', business feature, *M.E.E.D.*, 31 Aug. 1984, pp. 23–31.

Donald, D. (ed.), *The Pocket Guide to Airline Markings and Commercial Aircraft* (Temple Press/Aerospace, Feltham, 1985).

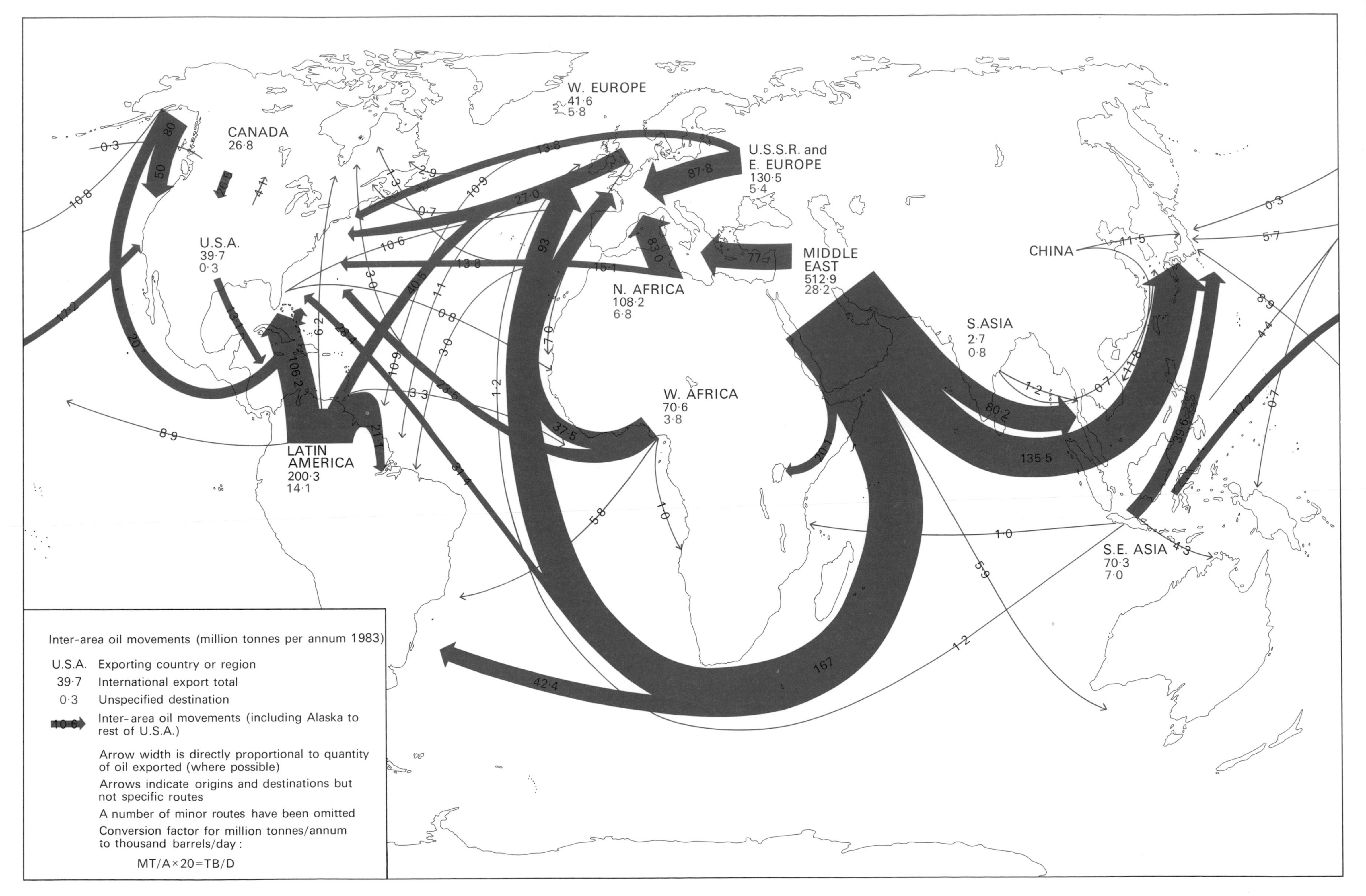

46 World oil movements

World oil movements

Gerald Blake

Total world energy consumption in 1984 was 7202 million tonnes of oil-equivalent, of which 39.5% was provided by oil. While world energy consumption is forecast to rise, the proportion contributed by oil is likely to decline somewhat; indeed, it has already declined from the peak year of 1979 when it provided 45% of the world's energy budget. Oil consumption is greatest in the United States, the Soviet Union, Japan and Western Europe (Table 20). The U.S.S.R. is currently self-sufficient for oil, but the United States, Japan and Western Europe import very large quantities, reflecting the boom decade of the 1960s when oil was a convenient and cheap energy source.

The Middle East possesses about 62% of the world's proven oil reserves, while local oil consumption is relatively small, being roughly equivalent to that of France or Italy. There are thus vast surpluses available for export, and world trade is dominated by movements from the Middle East to Western Europe and Japan. The North African countries of Libya, Algeria and Egypt are also significant producers and exporters of oil, amounting to 9% of the world oil trade in 1983, compared with 42.5% accounted for by Middle Eastern states. Thus North Africa and the Middle East together account for well over half the world's trade in oil. North African exporters have the undoubted advantage of being closer to their markets in Europe, with tanker voyages of a few hundred kilometres. By contrast, the distance from the Gulf to Japan is over 13,000 km and from the Gulf to Europe via the Cape route over 20,000 km. With such distances involved the economies of scale achieved by the use of supertankers are clearly important. Only a small proportion of world oil movements internationally are by pipeline. Over a given distance, movement by sea is cheaper and more convenient than by pipeline.

The map illustrates the world pattern of the combined movement of crude oil and oil products. In the early days of oil exports, relatively little was refined in the producer states and shipped as oil products. Middle East and North African states were slow to develop local refining capacity for a variety of reasons (Map 34). Today petroleum products make up about 27% of oil and product movements worldwide, but only account for about 10% of Middle East exports and 18% of North African exports. In 1983, 1206 million tonnes of oil and petroleum products entered world trade, twice as much as the combined volume of iron ore and coal. Thus crude oil is by far the most important seaborne cargo by volume. Not surprisingly, the construction and operation of oil tankers is big business. Although some producer states have attempted to transport oil in their own ships (Map 44) the world tanker fleet is still largely dominated by flag of convenience states (Panama, Liberia), by traditional seafaring states (Greece, Norway) and by the major importers (Japan, Western Europe). With the decline in oil consumption in the industrial countries, particularly since 1973, there has been a huge surplus tanker capacity and many relatively new ships have been scrapped or laid up. Nevertheless, oil tankers still make up about 35% of the world's merchant fleet by tonnage. More than half the total in 1983 comprised supertankers of over 250,000 gross tonnage. Some 16% were over 285,000 gross tonnage.

Oil-importing states are naturally very anxious about the security of their supply routes. Before 1967, attention in Western Europe was focused sharply on the Suez Canal, through which a high proportion of Middle East oil was shipped. Thus Britain and France invaded the Suez Canal zone in 1956 when they thought the security of the canal was threatened. After the closure of the Suez Canal in 1967 the Cape route was used, and in some quarters much was made of the possibility of interference with supertanker traffic rounding Africa, particularly in the Mozambique Channel and off the Cape of Good Hope. After the Iranian revolution and the Soviet Union's invasion of Afghanistan in 1979, the Strait of Hormuz was perceived as a potential choke point (Map 47), not least because Japanese as well as European and United States oil imports passed through. These events, coupled with the exigencies of the Iran–Iraq Gulf war, led to attempts to diversify oil export routes avoiding Hormuz. New pipelines are already in use, under construction, or planned (Map 33) and the Suez Canal has been widened to win back some of the oil trade from the Cape route (Map 50). Nevertheless, in 1983 over 80% of Middle East oil exports passed through Hormuz, more than one-third of which was destined for Japan. The Japanese thus watch the Gulf region very closely. They are equally concerned with the geopolitics of the region of the Strait of Malacca which constitutes another potential choke point for eastbound oil tankers. The strait is 930 km long, narrowing to 15.6 km. It is also shallow, averaging only about 27 metres. If the strait were closed to Japanese tankers, the alternative route to the south through Indonesian waters would be costly and not necessarily more secure. Although fears about physical threats to tanker routes have not been justified by events, they underlie much contemporary geopolitical thinking.

Table 20. *Oil and oil products: consumption and imports*

	Consumption 1983 (million tonnes)	Percentage of world consumption	Import (million tonnes)	Percentage imported
United States	704.9	25.5	262.4	37.2
Western Europe	586.6	20.8	427.9	72.9
U.S.S.R.	450.5	15.7	nil	nil
Japan	214.6	7.5	214.6	100.0

Source: British Petroleum, *Statistical Review of World Energy* (B.P., London, 1985).

KEY REFERENCES

British Petroleum, *Statistical Review of World Energy* (B.P., London, 1985).

Couper, A. D. (ed.), *The Times Atlas of the Oceans* (Times Books, London, 1983).

Drysdale, A. and Blake, G. H., *The Middle East and North Africa: A Political Geography* (Oxford University Press, New York, 1985).

O'Dell, P., *Oil and World Power*, 7th edn (Penguin Books, Harmondsworth, 1983).

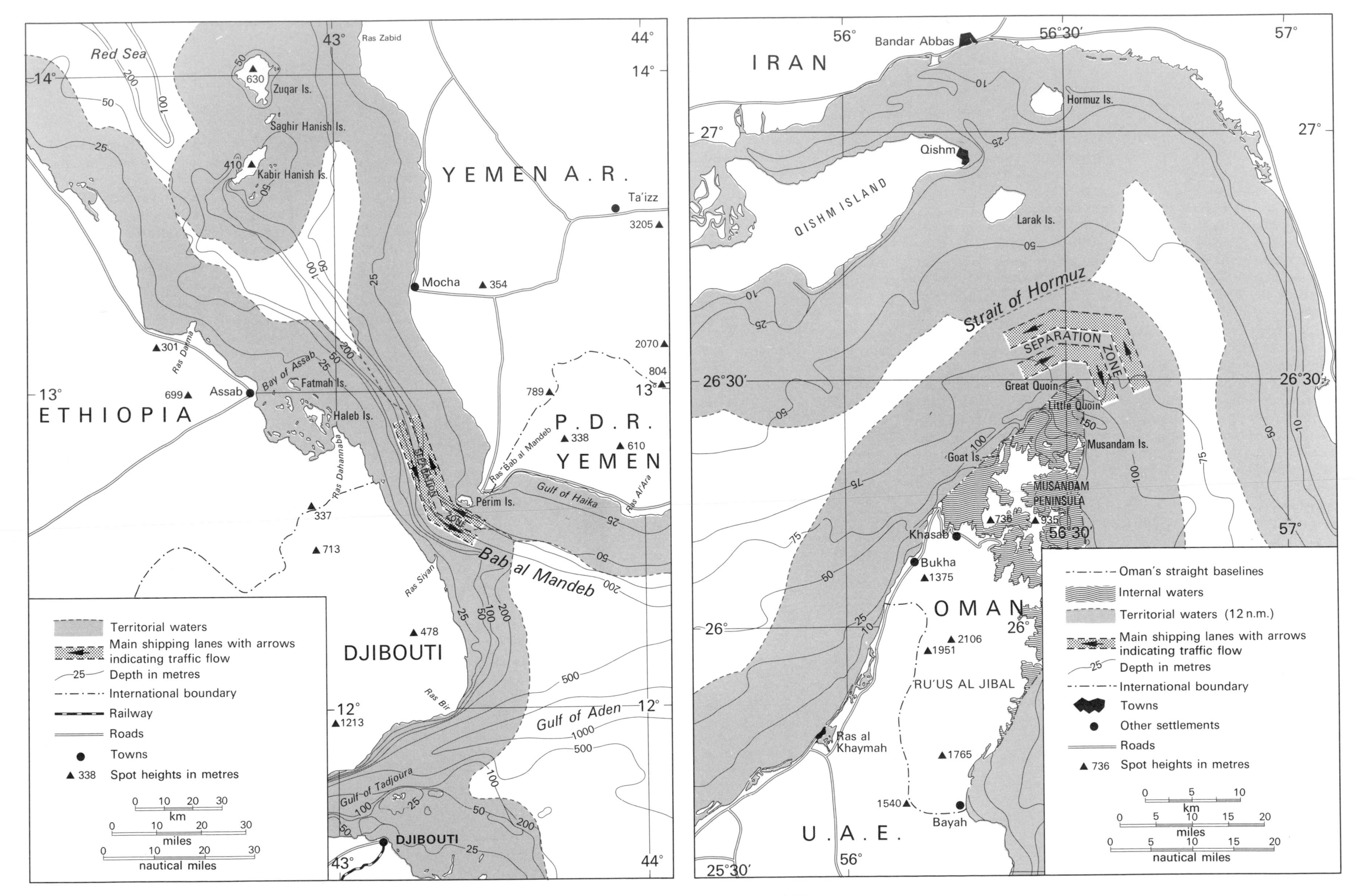

47 Bab al Mandeb and the Strait of Hormuz

Bab al Mandeb and the Strait of Hormuz

Gerald Blake

Bab al Mandeb

The continents of Africa and Asia are only 26 km apart at the Strait of Bab al Mandeb (Arabic for 'Gate of Lamentation'). The main shipping routes are west of Perim Island (P.D.R. Yemen), where the channel is 16 km wide and 100–200 m deep. There is thus no restriction on ship size, and it would be difficult to block such a deep channel. Bab al Mandeb is of great importance to shipping using the Red Sea and Suez Canal. At the same time it is the only natural outlet to the world's oceans for Jordan, Sudan, Yemen Arab Republic and Ethiopia, and is regarded as of great significance by Israel. Bab al Mandeb lies in the territorial waters of Yemen A.R., P.D.R. Yemen, Djibouti and Ethiopia. Of these, P.D.R. Yemen is best placed to exercise direct control over shipping lanes through its possession of Perim Island (8 km^2). Bab al Mandeb is the focus of considerable international attention because of its location in a region of great political instability and superpower rivalry. Soviet influence has been strong in Ethiopia since 1974 and in P.D.R. Yemen since 1979, while there is a French naval and military presence in Djibouti (population 200,000). Since 1980, the United States navy has used port facilities at Berbera in Somalia.

The number of ships using Bab al Mandeb is not recorded, but must be similar to Suez Canal transits. Thus allowing for Red Sea destinations, it is probably 20,000 ships annually or about 55 a day. Because so much of the traffic is associated with the Suez Canal, ships of many states and a wide variety of cargoes pass through Bab al Mandeb (Map 50). Before the 1967–75 closure of the Suez Canal it was the main sea route for the bulk of oil exported to western Europe and other markets. During closure, the Cape route came into prominence. Since 1975 the Suez Canal and the Suez–Mediterranean pipeline (Sumed) have recaptured some of the oil trade, but several new oil pipelines bypass Bab al Mandeb, and tanker traffic is no longer so significant (Map 33). The pattern could change markedly, however, if the Suez Canal were further widened to take larger supertankers. Bab al Mandeb is regarded as a vital waterway by diverse maritime interests, all of whom wish to see it remain open and secure for shipping. The 1982 United Nations Convention on the Law of the Sea introduced the concept of 'transit passage' in an attempt to prevent interference with shipping in straits by coastal states. The importance of this provision should also be noted in connection with the Strait of Hormuz (Map 47B) and Gibraltar (Map 51).

Strait of Hormuz

The Strait of Hormuz at the entrance to the Persian/Arabian Gulf has been the subject of more commentary over the past decade than any other strategic waterway. Since the Iranian revolution and the Soviet invasion of Afghanistan there have been widespread fears that Hormuz could be closed to shipping. These anxieties were most acute in the United States, Japan and Western Europe, who depended heavily upon imported oil passing through Hormuz. Although the threats to shipping may have been exaggerated they did have and still have some objective basis. The major Western industrialised countries import large quantities of Middle East oil (Map 46). Thus in 1983 76% of Japan's total crude oil imports, 49% of Western Europe's and 17% of the United States' crude oil imports came through the Strait of Hormuz. Because of a decline in oil consumption by the industrialised countries and some diversification of sources in recent years, these proportions are lower than at the end of the 1970s, when 90% of Japan's, 60% of Western Europe's and 30% of the United States' oil imports used Hormuz. Clearly, short-term interruption to these supplies would create inconvenience and higher prices, and serious economic damage in the long term. President Carter's administration declared that the United States was prepared to intervene in the Gulf region to secure supplies of oil to the Western world, and initiated the Rapid Deployment Force for this purpose. Hormuz is also vitally important to the Gulf states themselves, who depend on oil revenues, and import large quantities of industrial products and consumer goods including food.

Between 70 and 80 ships transit the Strait daily, including some of the largest supertankers afloat. The narrowest part lies between Larak Island (Iran) and the Quoin Islands (Oman), a distance of 39 km. Until 1979 tankers were permitted to use the passage between the Quoins and the mainland, in much more restricted waters. Since 1979 Oman has forbidden tankers to use this channel, which lies in her internal waters. Although the main navigation channels would be difficult to block physically, being 70–90 m deep, they lie close enough to the Musandam peninsula to be vulnerable to shore-based attack. Musandam is an arid, sparsely populated mountainous region fringed by a highly indented coastline reminiscent of the fiords of Norway. Although one of the most strategically important territories in the world, it could probably be seized and occupied by invading forces in a short time. In the age of Exocet missiles and the like it is clearly important who controls the coasts of strategic waterways. Formally, Oman and Iran are the riparian states. Since both claim territorial waters of 12 nautical miles their respective territorial waters are delimited by a median line for 28 km.

Various efforts have been made to ensure the safety of shipping in the Strait of Hormuz. In 1977 the Shah of Iran secured an agreement with Oman enabling his powerful navy to patrol tanker routes even in Omani waters. After the Iranian revolution in 1979 the Omanis appealed for financial and practical help in developing naval forces and bases to resume the task of patrolling the Strait. Today the Strait is regularly patrolled by units of the Omani navy, while the United States, United Kingdom and other powers maintain a naval presence in the area. The fact remains, however, that supertankers are vulnerable to attacks from the air, land and sea, and that their safety cannot be guaranteed. Thus alternative outlets for oil from the Gulf region are being sought via a series of new pipelines transporting oil direct to the Red Sea, Indian Ocean, and Mediterranean (Map 33). Such pipelines reduce dependence upon Hormuz chiefly for markets west of the Gulf region. Japan-bound exports, however, are less likely to find alternative routes, and Hormuz will continue to be the focus of much geopolitical activity.

KEY REFERENCES

Cordesman, A. H., *The Gulf and the Search for Strategic Stability* (Westview Press, Boulder, Colorado, and Mansell, London, 1984).

Lapidoth-Eschelbacher, R., *The Red Sea and the Gulf of Aden*, International Straits of the World 5 (Martinus Nijhoff, The Hague, 1982).

Ramazani, R. K., *The Persian Gulf and the Strait of Hormuz*, International Straits of the World 3 (Martinus Nijhoff, The Hague, 1979).

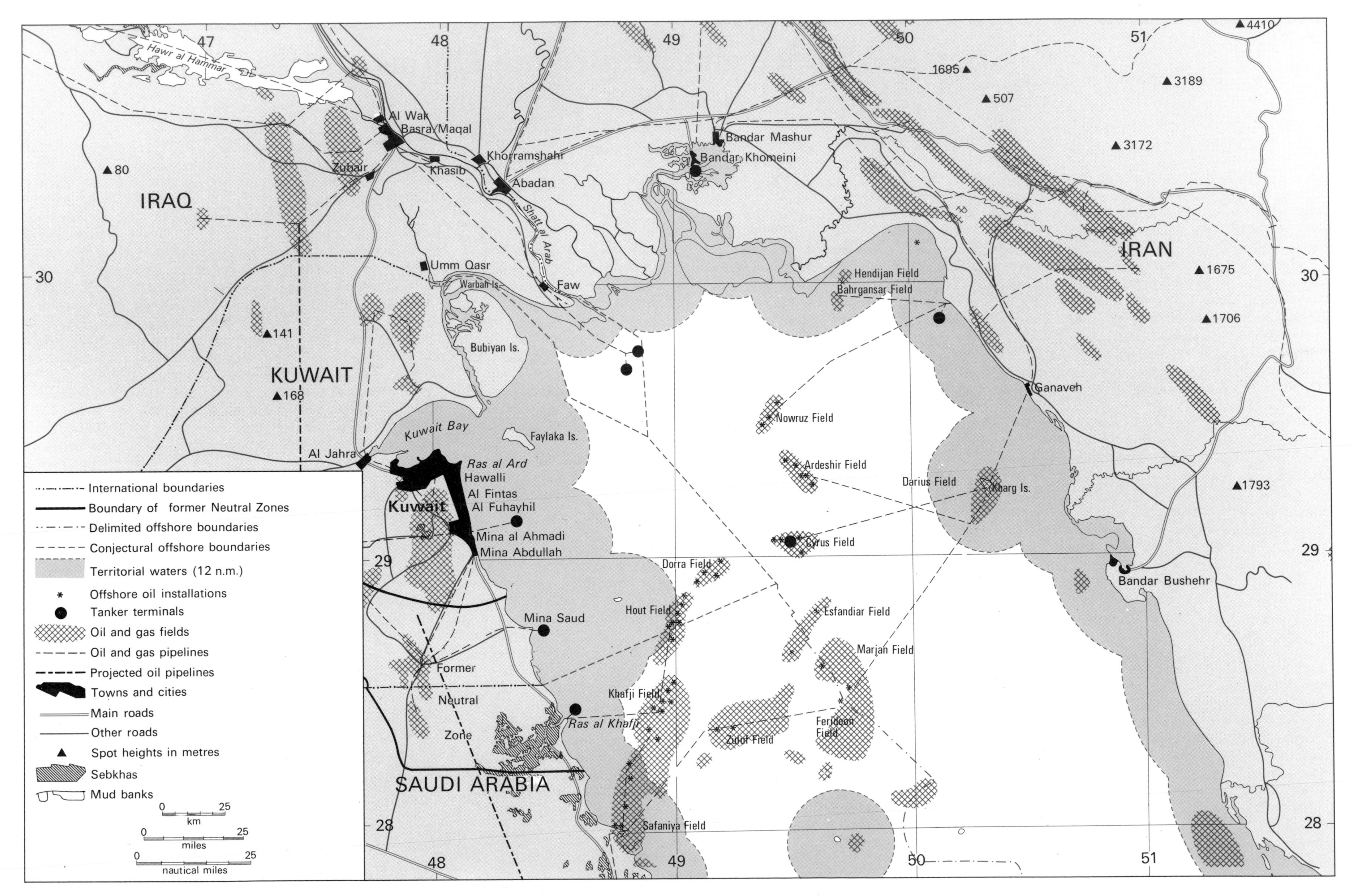

48 The Upper Gulf Region

The Upper Gulf region

Gerald Blake

Four states (Kuwait, Saudi Arabia, Iraq and Iran) border the waters of the northern Gulf, two of which (Kuwait and Iraq) have no other direct access to the sea. The geopolitical importance of the Upper Gulf today rests primarily upon its role as an oil-exporting region, and as a zone of contact between the Arab and Persian worlds. This ancient cultural cleavage, which runs more or less through the Upper Gulf, has received international attention following the outbreak of the Iran–Iraq war in 1980. The Upper Gulf was also the focus for international interest in the late nineteenth century, when German proposals to build a strategic railway from Europe to the Gulf were seen as a threat to Britain's Gulf route to India. The term 'Middle East' probably entered common parlance at that time to denote the region surrounding the Gulf. One upshot of the Anglo-German rivalry was Britain's willingness to establish a protectorate over Kuwait when invited to do so by Sheikh Mubarak in 1899. British protection may have helped Kuwait remain independent in the 1920s, when the Saudis were expanding and consolidating their territorial gains in the Arabian peninsula. The British protectorate ended in 1961. Iraq at once claimed Kuwait on the grounds that it had formerly been part of the Ottoman province of Basra. While this claim was dropped in 1963, Iraq has persisted in claiming the low-lying, uninhabited islands of Warbah and Bubiyan, whose possession would improve Iraq's access to the port at Umm Qasr and entitle Iraq to a greater share of the seabed. In 1975 Kuwait built a 4 km road bridge to Bubiyan as a way of reinforcing control of the island.

International boundary delimitation has posed a number of problems in the Upper Gulf on land and sea. When the Kuwait–Saudi Arabia boundary was agreed in 1922, a Neutral Zone (5700 km^2) was left in a region whose ownership was undecided. In 1966, this Neutral Zone was partitioned between the two states, although revenues from oil and other natural resources are shared by Kuwait and Saudi Arabia. The partition agreement did not resolve disputed ownership of the islands of Qaru and Umm al Maradin which lie off the former Neutral Zone. The islands were allegedly occupied by Saudi Arabia in 1977, but are claimed by Kuwait. Although the islands are small, large areas of seabed are involved. Following border agreements in 1981 the Neutral Zone west of Kuwait was partitioned between Saudi Arabia and Iraq. The Kuwait–Iraq boundary was also delimited in 1922 and has been the scene of many border incidents, although the Iraqis do not seem to question its general alignment.

The position of the boundary between Iraq and Iran along the river Shatt al Arab was one of the immediate causes of war in 1980. An Ottoman–Persian agreement in 1847, confirmed in 1913, gave Iraq control of the river (Fig. 6). In 1937 the boundary opposite Abadan was moved to the *thalweg* or deepest point (A to B on Fig. 6). Iranian demands for a *thalweg* boundary throughout were conceded by Iraq at Alger in 1975 in return for the withdrawal of Iranian support for Kurdish rebels in northern Iraq. Iraq abrogated the Alger agreement in 1980, when relationships with the revolutionary regime in Iran were fast sliding towards war. The outcome of the Gulf war and its long-term political and economic consequences have yet to be seen. Meanwhile, the vulnerability of offshore oil installations which are especially concentrated in the Upper Gulf has been amply demonstrated. Damage caused to rigs in the Nowruz and Ardeshir fields in 1983 caused oil leakages which created costly pollution in many parts of the north and central Gulf.

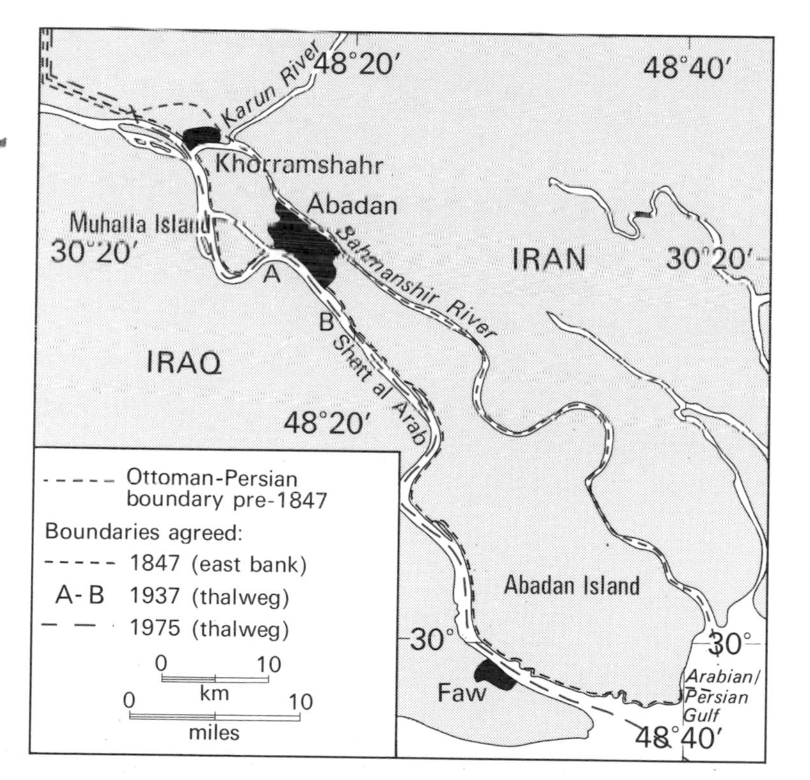

Fig. 6 Iraq–Iran boundary along the Shatt al-Arab.

Several oilfields lie offshore in the Upper Gulf. Being located in shallow water rarely exceeding 60 m and relatively close to the shore, they are clearly attractive to exploit (Map 33). Unfortunately, exploration and development have been held up in certain areas by disputes over boundary location. Only the Saudi Arabia–Iran boundary has been formally agreed, in 1968. Disputes over ownership of islands (see above) account for problems between Saudi Arabia and Kuwait, and Iraq and Kuwait. The main difficulty in the allocation of continental shelf between Kuwait and Iran is the presence of offshore islands whose role in boundary delimitation cannot be agreed. A line based on mainlands would favour Iran and one based on islands would favour Kuwait.

The Upper Gulf states all claim territorial waters of 12 nautical miles (22.24 km). The effects of offshore islands can be clearly seen, since islands are entitled to their own territorial waters. With no offshore islands and a coastline of only 19 km, Iraq is particularly disadvantaged with regard to maritime access and resources.

KEY REFERENCES

Cordesman, A. J., *The Gulf and the Search for Strategic Stability* (Westview Press, Boulder, Colorado, and Mansell, London, 1984).

Cottrell, A. J. (ed.), *The Persian Gulf States* (Johns Hopkins University Press, Baltimore and London, 1980).

McLachlan, K. and Joffe, G., *The Gulf War*, The Economist Intelligence Unit Special Report no. 176 (Economist Publications, London, 1984)

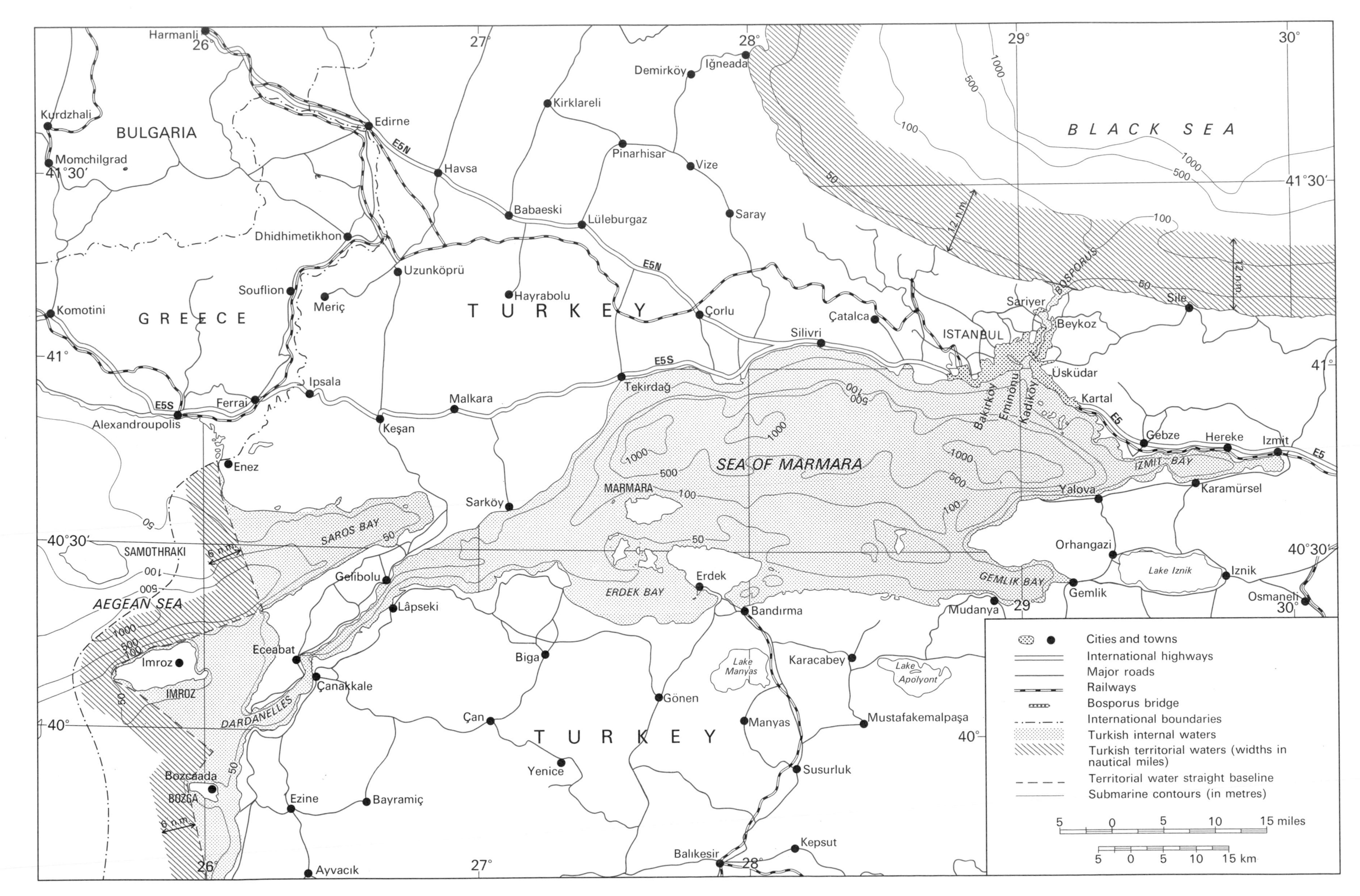

49 The Turkish Straits

The Turkish Straits

Gerald Blake

The Turkish Straits differ from the other strategic waterways affording access to the Middle East and North Africa in several important respects (see Maps 47 and 51). Both shores of the straits are in the hands of a single state, and control is easily exercised because of their length and narrowness. Their strategic significance is quite unequivocal, since they give the Soviet Union's powerful Black Sea fleet access to the Mediterranean and beyond. Turkey is a member of the N.A.T.O. alliance and the region of the Turkish Straits is thus one of the most sensitive contact zones between the Warsaw Pact and N.A.T.O. The straits are also of considerable domestic importance to Turkey; indeed, the great city of Istanbul (population 5.7 million) straddles the Bosporus. With a large resident coastal population and concentrations of heavy industry, parts of the Bosporus and Sea of Marmara are badly polluted.

The Turkish straits comprise the Bosporus and Dardanelles, whose importance to sailors has been recognised since the Trojan wars 3000 years ago. The Bosporus is about 27 km long and only 640 m wide at its narrowest point. Though the main channel is deep (70 m) navigation is made hazardous by fast currents and sharp bends. Numerous accidents to shipping and even houses on the shore have occurred, and the Turkish authorities are always anxious about the dangers of pollution. The Dardanelles are 58 km long, narrowing to 868 m. Depths vary from 50 to 90 m. The two straits are separated by the Sea of Marmara, whose length of over 200 km represents several hours of navigation in Turkish waters for ships in transit. The Dardanelles are more important for Turkey's overseas trade than the Bosporus, which is the major crossing point between Europe and Asia. A high-level bridge over the Bosporus was completed in 1973, and construction on a second bridge began in 1986 to cater for the great increases in local and international traffic.

The question of rights of transit through the Dardanelles and Bosporus goes back centuries. In the fifteenth century the Ottomans prohibited the passage without permission of foreign ships. This remained unchallenged until the Russians gained a foothold on the northern shores of the Black Sea, and earned the right of unimpeded navigation for merchant ships in the Treaty of Kuchuk Kainarji (1774). Under the Treaty of Paris (1856) after the Crimean War, commercial navigation was open to all though passage of warships was still restricted. After World War I, the straits region was demilitarised, and administered by an International Straits Commission (Treaty of Lausanne 1923). These arrangements were superseded by the Montreux Convention of 20 July 1936 which was signed by ten states (Bulgaria, France, Great Britain, Australia, Greece, Japan, Romania, Turkey, Yugoslavia and the U.S.S.R.) and remains in force today. Basically, its 29 articles regulate the passage of warships in and out of the Black Sea, and make Turkey gatekeeper of the straits. Although the Russians have generally been careful to abide by the Montreux Convention there is no doubt they would like to see it changed. Concessions were actively sought in 1914 and 1945 but came to nothing. Although the Soviet Union's interests in the Turkish straits are both commercial and strategic, it is the movement of warships in peacetime which are most affected by the Convention. In wartime there can be little doubt that Soviet forces would be obliged to try to occupy the straits region to ensure access to the Black Sea. In peacetime, at least eight days' prior warning of the passage of warships has to be given to Turkey, with details of the ship and its destination. No more than nine warships may transit at any time, and their aggregate tonnage must not exceed 15,000 tons, although an exception is made for capital ships of over 15,000 tons. Transit must be in daylight, while submarines must transit on the surface, and singly. Submarines are not supposed to use the straits for operational purposes, but only for repair and maintenance, and the passage of aircraft carriers is forbidden. There are also clauses to ensure that non-Black Sea naval forces do not pose a threat to Black Sea powers. Turkey has the right to prevent passage of warships altogether if national security is threatened.

Table 21. *Merchant ships in transit through the Bosporus*

	1981		1982		1983	
	Ships	Gross tonnage (millions)	Ships	Gross tonnage (millions)	Ships	Gross tonnage (millions)
U.S.S.R.	3,360	36.8	6,054	65.0	4,793	56.6
Greece	2,316	21.8	1,727	19.8	1,302	18.3
Romania	1,533	18.3	1,672	23.2	1,695	21.3
Others	4,313	37.5	4,594	36.0	4,832	44.3
Total	11,522	114.4	14,047	144.0	12,622	140.5

Source: G. H. Blake, 'Marine policy issues for Turkey', *Marine Policy Reports*, 7 (4), 1985, pp. 1–6.

The crucial importance of the Turkish straits to the Soviet navy is related to the fact that none of the Soviet Union's four fleets enjoys unimpeded access to the world's oceans. The Black Sea fleet, consisting of some 85 surface combat ships and 25 submarines, provides units for operation in the sensitive waters of the Mediterranean and Middle East, and occasionally for the eastern Atlantic. There are normally 40 to 50 Soviet ships in the Mediterranean. In the early 1960s only about 30 Soviet warships passed through the straits annually, but by 1985 more like 250 transits were made. Warships of other states rarely use the straits, apart from the United States navy, which regularly patrols the Black Sea.

Merchant ships may use the Turkish Straits without restriction, apart from normal sanitary controls. No dues are charged and pilotage is not obligatory, though strongly recommended. About 28,000 ships pass the Turkish straits in a year, including much Turkish coastal shipping. This represents more than 70 ships a day, a figure comparable with the Suez Canal, though the average tonnage of ships using the canal is about twice as great. The majority of ships transiting the Turkish Straits are commercial. The number of Soviet merchant ships in transit has doubled in the last 20 years, and by 1983 accounted for about one-third of the total number, and about 40% of gross tonnage (Table 21). This proportion seems likely to go on increasing in future, since the Black Sea ports serve the Soviet Union's industrial heartland, linked to an extensive network of navigable rivers and canals. Greece and Romania are other important commercial users of the straits. All kinds of ships and cargoes are involved, including cruise ships in surprising numbers, and cargoes of grain, chemicals and oil. The possibility of closure of the Turkish Straits is not considered to be very great. Certainly Turkey has every interest in keeping it open. Nevertheless, conflict in the Aegean between Greece and Turkey, if it ever occurred, might result in dangers to shipping in the Turkish Straits. Turkey is extremely sensitive about maintaining high seas approaches to the entrance to the straits in the Aegean Sea, rather than letting them fall into Greece's territorial seas. Turkey therefore maintains a 6 nautical miles territorial sea claim in the Aegean, but 12 nautical miles in the Black Sea and Mediterranean.

KEY REFERENCES

Vali, Ferenc, *The Turkish Straits and NATO* (Hoover Institution Press, Stanford University, Stanford, 1972).

The Suez Canal

Gerald Blake

The idea of a canal across the narrow Isthmus of Suez goes back to 2000 B.C. The earliest canal appears to have made use of a natural wadi running westwards to the Nile delta from Lake Timsah to facilitate trade between the delta and the Red Sea. The ancient canal was eventually left to silt up. Although Napoleon Bonaparte was interested in the prospects for a modern canal, no action was taken until 1854 when a French engineer, Ferdinand de Lesseps, was granted permission to construct a canal by the Khedive Muhammad Said. Operating rights for 99 years were granted to the Compagnie Universelle du Canal Maritime de Suez in 1856. The route chosen by de Lesseps was not the shortest, but made use of several natural lakes and marshes (see map). The canal opened in 1869 and thereafter had a marked impact on world shipping, particularly affecting British trade. Indeed, its importance to the oil-importing states of Europe became so great that Britain and France were prepared to go to war in 1956 to protect their interests, following nationalisation of the canal company by President Nasser of Egypt. As a result of hostilities the canal was blocked by shipwrecks until April 1957. After a further decade of growth in traffic, especially oil cargoes, the canal was again closed to shipping after Israel invaded Sinai in June 1967. It did not reopen until June 1975. During the eight years of closure, international oil transportation had been revolutionised by supertankers using the Cape route. By June 1975 only 27% of the world's tankers could negotiate the canal.

The Suez Canal provides a short cut between the Mediterranean and Red Seas, and between the Atlantic and Indian Oceans beyond. Using the canal can save vessels from 10% to 50% in distance and from 50% to 70% in fuel consumption, depending on ship speed, size and destination (Table 22). While the canal's importance for oil transportation has declined, its significance for world trade remains

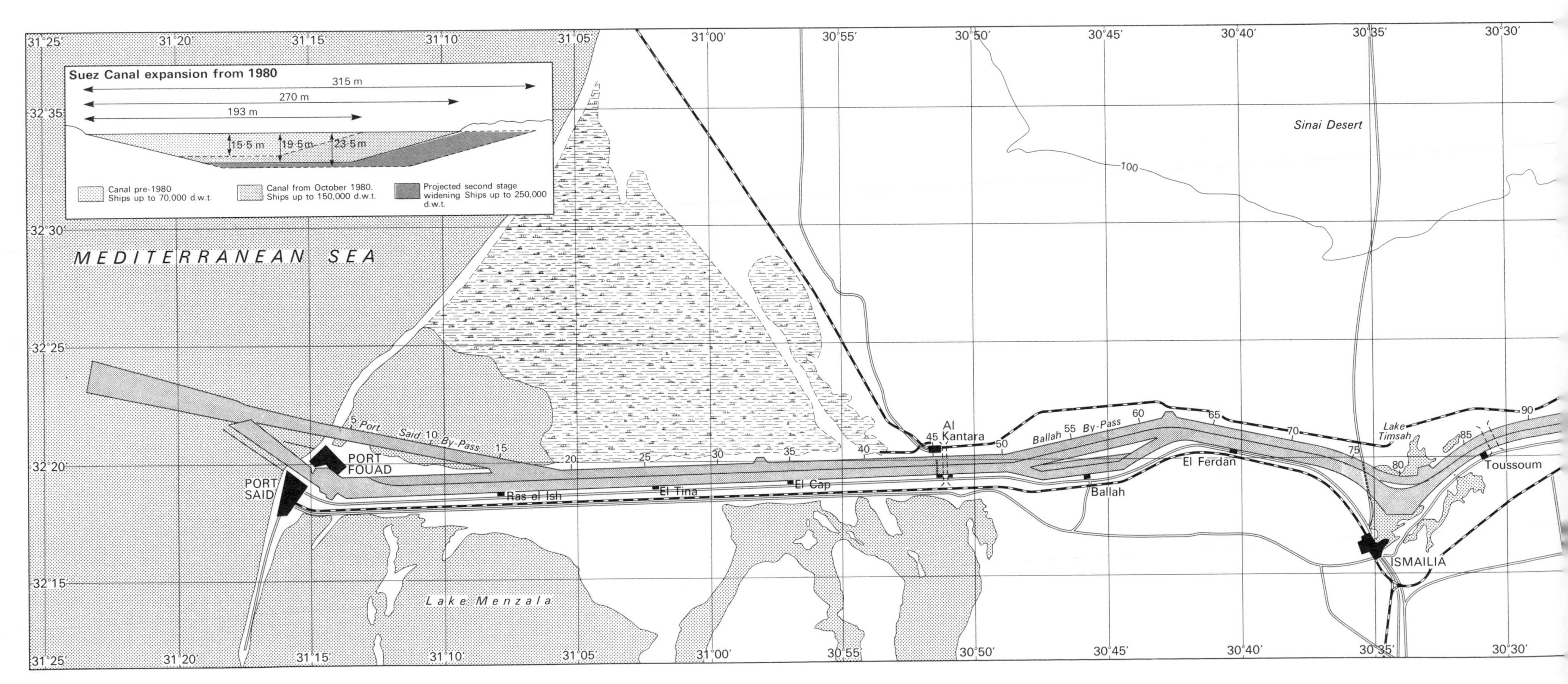

50 The Suez Canal

considerable. About 7% of world seaborne trade transits the canal, compared with about 15% in the 1960s. Because of the enormous

Table 22. *Distances between selected ports (km)*

	Via the Cape	Via Suez Canal	Distance saving (%)
London to:			
Bombay	20,000	11,667	42
Kuwait	20,928	12,039	42
Melbourne	22,594	20,372	10
Singapore	21,852	15,370	30
New York to:			
Bombay	21,854	15,185	31
Singapore	23,150	18,890	18
Ras Tannurah	22,038	15,370	30

Source: Adapted from W. B. Fisher, 'The Suez Canal', *Encyclopaedia Britannica*, 15th edn (Encyclopaedia Britannica, Chicago, 1974), p. 768.

increases in the volume of world trade, 7% represents approximately the same volume of goods as in the 1960s. Thus 264 million tons of goods passed through the canal in 1984, compared with 274 million in 1966, the peak year before closure. Cargoes include considerably less oil than in 1966, but oil and oil products still account for some 37% of all goods by volume. The chief commodities southbound are fertilisers, cement, petroleum products, and metal and manufactured goods. Northbound commodities include oil and petroleum products (56% by volume), ores and metals, oil seeds, textile fibres, and other raw materials. The largest northbound oil cargoes go to southern Europe (57% to Italy), northern and western Europe (10%), and North America (10%).

There have been many enlargement and improvement schemes since the surprisingly small original canal, which could take ships of maximum draught of 24 ft. Eight major schemes were completed before 1967, enabling ships of 38 ft draught (70,000 d.w.t. laden) to pass through. After reopening in 1975, a major canal development

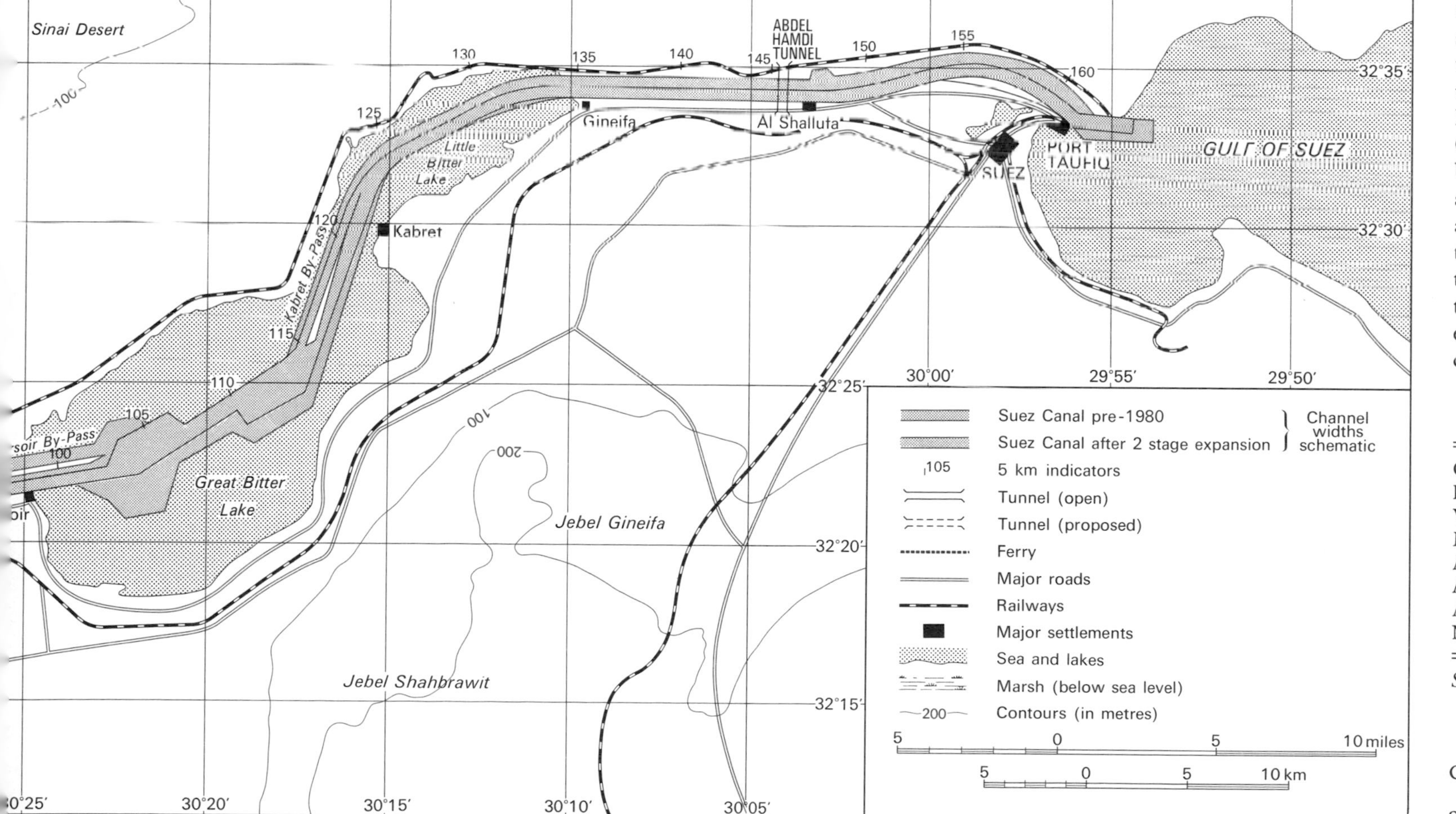

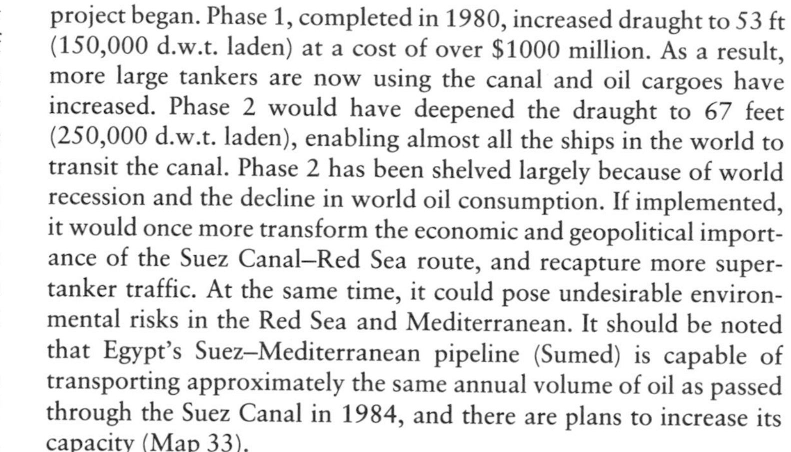
project began. Phase 1, completed in 1980, increased draught to 53 ft (150,000 d.w.t. laden) at a cost of over $1000 million. As a result, more large tankers are now using the canal and oil cargoes have increased. Phase 2 would have deepened the draught to 67 feet (250,000 d.w.t. laden), enabling almost all the ships in the world to transit the canal. Phase 2 has been shelved largely because of world recession and the decline in world oil consumption. If implemented, it would once more transform the economic and geopolitical importance of the Suez Canal–Red Sea route, and recapture more supertanker traffic. At the same time, it could pose undesirable environmental risks in the Red Sea and Mediterranean. It should be noted that Egypt's Suez–Mediterranean pipeline (Sumed) is capable of transporting approximately the same annual volume of oil as passed through the Suez Canal in 1984, and there are plans to increase its capacity (Map 33).

Trade to and from almost every corner of the world passes through the Suez Canal, though the volumes vary greatly. Latin America and Australasia use the canal very little. At the other extreme, over 30% of goods imported into the Mediterranean and 17% of exports transit the canal. About 15% of goods both exported and imported by northern and western Europe also use the canal. The Arabian peninsula states and the Red Sea littorals are similarly highly dependent upon canal transportation. The top half dozen flags measured by total ship tonnage include Liberia (11.7% in 1984), Greece (10.6%), Panama (7.7%), U.S.S.R. (5.9%) and the United Kingdom and Japan. Counting numbers of transits, China and West Germany displace the United Kingdom and Japan. Overall, however, the canal is incomparably most significant to Egypt as a major earner of foreign exchange, and the *raison d'être* for the only major urban development outside the Nile valley and delta. Canal defence is a top strategic priority, and three road tunnels are being constructed under the canal to facilitate the defence of Sinai and the canal's eastern flank. A free zone has been established with some success at Port Said to stimulate industrial and commercial activity based on the canal's favourable global location.

Table 23. *Suez Canal: facts and figures*

Overall length Port Said to Port Taufig	193.5 km (120.2 miles)
Port Said to Ismailia	78.5 km (48.8 miles)
Width	300–350 m (328–383 yards)
Maximum speed for ships in transit	14 km (8.7 miles)/hour
Average transit time	24 hours
Average time spent in passage	12 hours
Average daily transits (1984)	58
Number of flags transiting (1984)	95

Source: Suez Canal Authority, *Yearly Report 1984* (S.C.A., Ismailia, 1985).

KEY REFERENCES

Couper, A. D. (ed.), *The Times Atlas of the Oceans* (Times Books, London, 1983).
Suez Canal Authority, *Yearly Report* (S.C.A., Ismailia, annual).

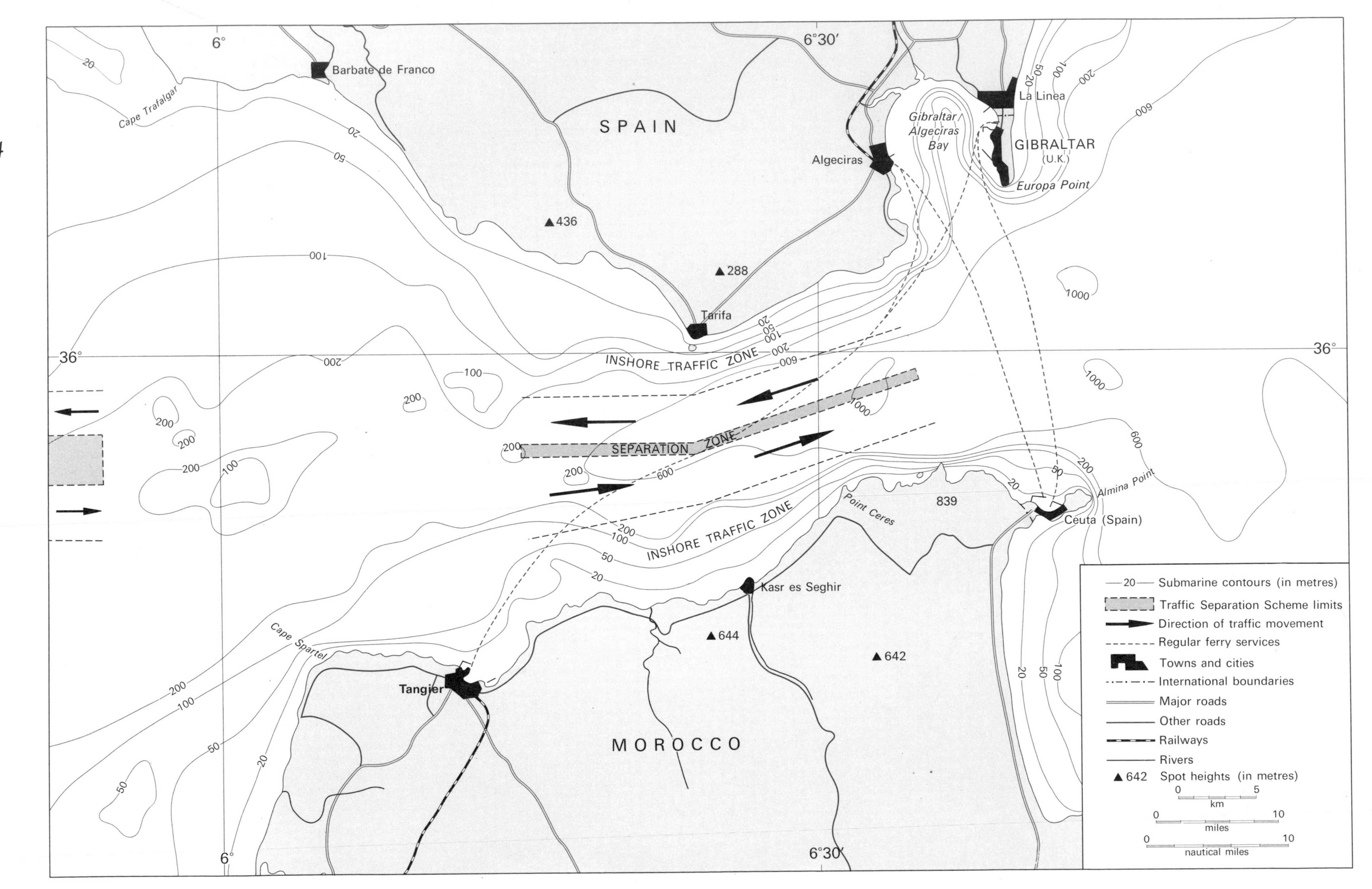

51 The Strait of Gibraltar

Strait of Gibraltar

John O'Reilly

The Strait of Gibraltar is the only natural entrance to the semi-enclosed Mediterranean Sea, and is one of the world's most vital shipping routes. It is approximately 58 km long, narrowing to 12.5 km. The maximum water depth in the main channels used for shipping is 935 m and the minimum depth is 320 m. The Strait plays a crucial physical role in the Mediterranean Sea; about 75% of Mediterranean water lost by evaporation is replaced by inflowing Atlantic currents. The salinity level is 38 parts per thousand before the Mediterranean water is diluted by Atlantic waters. Mediterranean water turnover via the Strait takes 70 to 80 years, a key factor in the light of high pollution levels. The Vandavals, Levanter and Sirocco winds influence conditions significantly, producing the infamous mists and dust clouds which may reduce visibility to less than 5 nautical miles. Shipping disasters were common in the Strait in the past.

The Strait of Gibraltar is probably the second busiest inter-oceanic strait after the Strait of Dover. Approximately 10 to 12 vessels can be observed in the Strait at any time of day. More than 150 vessels of over 1000 gross tonnage transit daily, a third of which are oil tankers, or approximately 73,000 ships a year (excluding small vessels and submarines). It is estimated that at least 50 million tonnes of crude oil were transported via the Strait in 1982, equivalent to 5% of global oil movements by sea. Besides oil, phosphates, iron ore, liquefied natural gas, aluminium bauxite and grain pass through, mostly northbound. Southbound manufactured goods are also important. Since the Camp David Agreement (1978) and the widening of the Suez Canal in 1980, east–west traffic has increased. There are also up to a dozen ferry-container services daily between Tangier and towns on the European shore, and hourly daylight crossings between Algeciras and Ceuta. Risk of collision is considerable with such intense traffic, and coastal states are particularly worried about the danger of marine pollution. An I.M.O. (International Maritime Organisation) traffic separation scheme is now in operation. Possible dangers to shipping have been a major consideration in the choice of a fixed link across the straits. The Moroccans originally favoured a pontoon and suspension bridge combination, while a causeway and bridge was at one time suggested by Spain. However, despite considerable technical difficulties, produced as much as anything by the inconsistent depth of the straits, a rail tunnel was agreed upon in principle in 1982. This would provide an invaluable link between the European and North African transport systems.

Every leading maritime power has striven to control the Strait, among them the Phoenicians (1100 B.C.), Greeks (700 B.C.), Carthaginians (600 B.C.) and Romans (200 B.C.). The Arab Tarek-el-Zaid led the *jihad* across the Strait in A.D. 711; 'Gibraltar' is a corruption of the Arabic 'Djebel Tarek' or 'Tarek's Mountain', denoting the famous 'Rock'. The Moors banned non-Muslim transit through the Strait, but the north–south axis acted as a bridge for the flow of people, goods and ideas. The Great Sahel–Mediterranean Gold Route flourished from A.D. 1100–1400. The Spanish retook the northern shore from the Moors in 1502 and continued the crusades into Africa, establishing 'Plazas' in Ceuta (1580), Melilla (1497), Alhucemas (1673), Velez de la Gomera (1508) and Chaffarinas Islands (1848), which remain Spanish sovereign territory to this day. These enclaves could entitle Spain to a generous share of the seabed under the United Nations Convention on the Law of the Sea of 1982. A combined Anglo-Dutch force took the Gibraltar peninsula in 1704, and British sovereignty over Gibraltar was established by the Treaty of Utrecht (1713). The 'Rock' has been besieged by Spain many times, notably during the Great Siege (1779–83). A successful economic blockade was also imposed by Spain from 1969 to 1985. With the rise of the British Empire and the opening of the Suez Canal (1869), Gibraltar became one of Britain's most valuable strategic assets. In the nineteenth century, Spain, France and Britain vied for control of the Strait, or at least tried to prevent rivals from exercising control. Hence Britain's intervention in the Spanish–Moroccan War (1859–60). The international treaty of 1865 establishing the 'special status' of the Cape Spartel lighthouse was the first attempt at internationalising the Strait. The Anglo-French Declaration of 1904 confirmed France's interests in Morocco. The Algeciras Act (1906), signed by ten states, while reasserting Moroccan sovereignty paved the way for the establishment of the Franco-Spanish Protectorate (1912–56). The erection of fortifications and strategic works on the southern shore was forbidden but neither demilitarisation nor proper neutralisation was implemented. The nearest the Powers came to internationalising the Strait was the establishment of the Tangier Neutral Zone which lasted from 1923 to 1959. In the 1950s, the United States established its largest relay station there, transmitting the 'Voice of America'. Because of Spanish neutrality during the world wars, the British possession of Gibraltar proved invaluable, in denying passage to enemy shipping, as an assembly point for convoys, and in victualling, refitting and repairing ships. It served as the key assembly point for the Allied invasion of North Africa (Operation Torch, 1943), and was also used during the Falklands/Malvinas War (1982). Today the Rock provides a base for British and N.A.T.O. naval units and Signal Hill serves as a ship monitoring station. With the advent of modern weapons and Spain's entry into N.A.T.O., Gibraltar's strategic importance to Britain is much diminished. But it should be noted that the Mediterranean is an extremely important theatre for submarines. The United States and other N.A.T.O. nuclear-powered missile submarines use the Strait as do Soviet submarines which cannot traverse the Turkish Straits for operational purposes under the Montreux Convention (1936). Gibraltar Strait is also vital to United States allies in the region, such as Israel. Geostrategically, Gibraltar and Malacca may be considered to be the most important straits in the world. By almost any criteria the Strait of Gibraltar ranks high, reckoned by number of ships, nationalities transiting, military uses, transportation of energy and vital raw materials and geographical vulnerability. Unjustifiably, its importance has often been overshadowed by Bab al Mandeb and Hormuz in the Middle East (Map 47).

The Strait of Gibraltar may be poised to become the foremost link between Africa and Europe, while east–west shipping is increasing steadily. The main service and transit ports in the area are Gibraltar (population 29,000), Ceuta (80,000), Algeciras (100,000), and Tangier (250,000), the two former being 'free ports'. The contentious questions concerning enclaves and territorial waters have yet to be settled, while the more acute socio-economic and political problems on the southern shore could also give rise to unrest, resulting in threats to shipping in the Strait, and rendering it a potential flash-point.

KEY REFERENCES

Hydrographer of the Navy, *Mediterranean Sea Pilot*, vol. 3 (Hydrographic Dept, M.O.D., Taunton, 1978).

Lapidoth-Eschelbacher, R., *Les Detroits en droit international* (Pèdone, Paris, 1972).

Truver, S. C., *The Strait of Gibraltar and the Mediterranean*, International Straits of the World 4 (Martinus Nijhoff, The Hague, 1980).

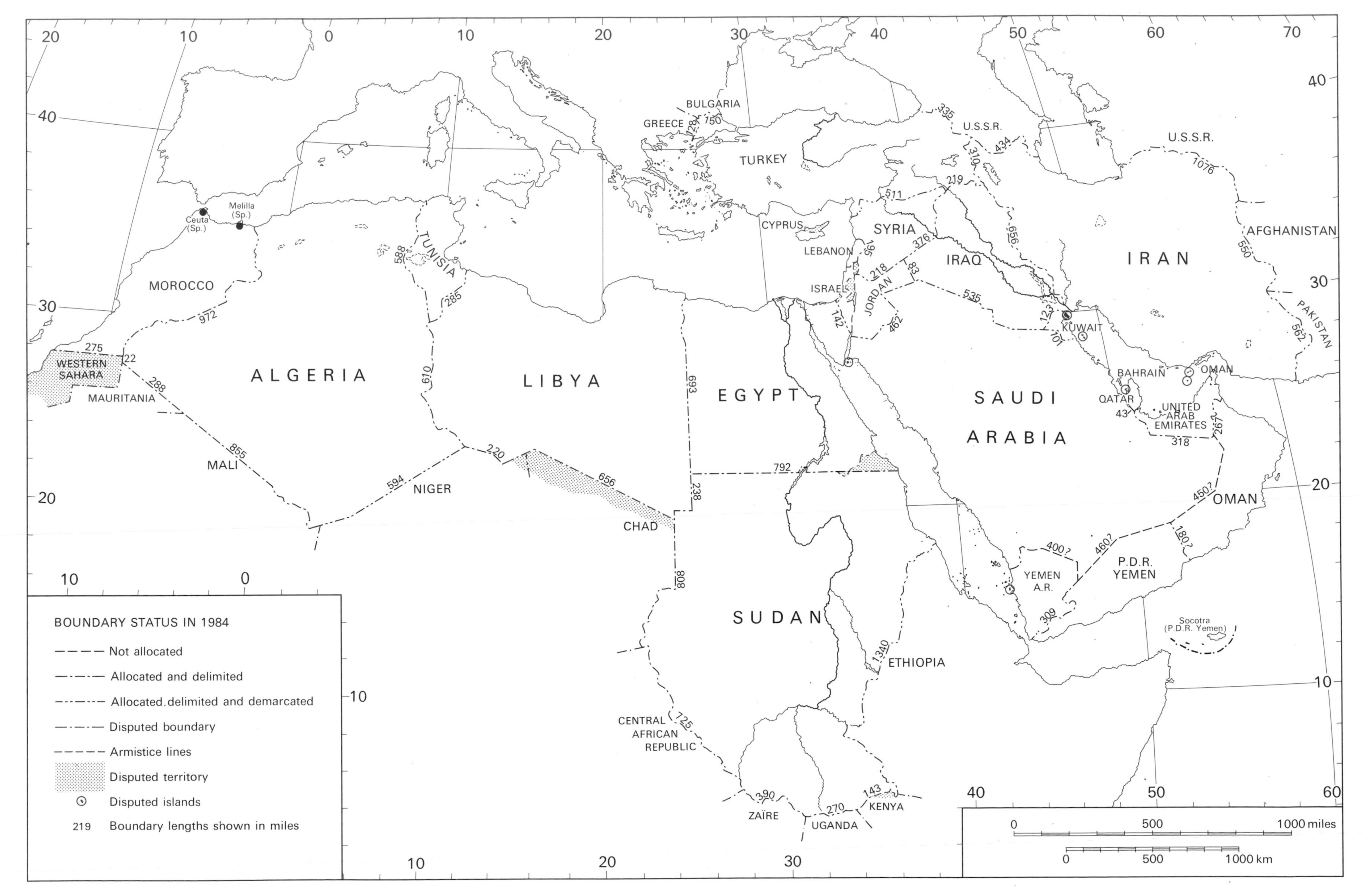

52 Inter-state land boundaries

Inter-state land boundaries

Gerald Blake

The international land boundaries which outline the political map today are relatively recent creations. The majority were aligned between about 1880 and 1930, largely as a result of European intervention. Some roughly coincide with cultural and historic divides, but the majority were superimposed with scant regard for the human or physical geography of the region. Boundaries act as filters to the flow of people and goods, and the closure of international boundary crossings has been common at times of political tension in the Middle East and North Africa. Some of the world's least permeable boundaries are in the Middle East, for example between Turkey and the Soviet Union and Syria and Israel. On the other hand, the great length of many boundaries, often passing through sparsely inhabited regions, leaves ample scope for illegal border crossing.

There are over 34,000 km of boundary shown on the map, varying considerably in legal status and physical characteristics. Saudi Arabia's boundaries with her three southern neighbours have not yet been aligned in what is largely uninhabited desert and the lines shown are hypothetical. Everywhere else, boundaries between states have been delimited, although one or two territorial disputes could result in future boundary changes. Surprisingly, perhaps, over 13,300 km of boundary are demarcated on the ground by physical or man-made features. In certain desert areas, notably in the Sahara, artificial boundary demarcations using markers of some kind would be difficult and costly. Long, straight boundaries are a prominent feature of the map. In North Africa these reflect delimitations agreed by the imperial powers, chiefly Britain and France, during the nineteenth-century scramble for Africa, while in the central Middle East they are the product of the post-World War I peace settlement.

International land boundaries can be classified according to their physical characteristics, whether they follow physical features, man-made features, or geometric guidelines such as lines of latitude or longitude (see Table 24). All boundary lines are artificial creations, subdividing river basins, cutting off towns from their natural hinterlands, and interrupting nomadic movements and communications. In these respects, Middle East and North African boundaries are no better and probably no worse than boundaries elsewhere.

Several territorial disputes remain unsettled, in addition to disputed sovereignty over a number of offshore islands (see map). Morocco claims Western Sahara (formerly Spanish Sahara) on historic grounds. When Spain withdrew in 1976 the territory was partitioned between Mauritania and Morocco but Mauritania gave up the southern half in 1978. Moroccan control is fiercely contested by the Polisario Front, who want independence. Morocco would also like to see an end to Spanish possession of the small urban enclaves of Ceuta and Melilla on the Mediterranean coast. Libya occupied large parts of northern Chad (the Aouzou strip) in 1975 on the grounds that this territory was ceded by the Italian administration to the French in 1935, although the treaty was never implemented.

Turkish forces invaded Cyprus in 1974 to protect the interests of the Turkish Cypriot minority. *De facto* partition of the island followed, with the Turkish community holding roughly 37% of the land, and the Greek Cypriots roughly 63%. Israel and the Arab territories occupied by Israel in 1967 are also shown on the map as disputed territory. The Palestinians claim Palestine as their own, and repudiate the creation of Israel in 1948. Syria insists on the return of the Golan region occupied in 1967 and Jordan equally demands the return of the West Bank to Arab control. Territorial difficulties also exist over the Sudan–Egypt boundary. Both states claim a small piece of territory (the Wadi Halfa salient) now under the waters of Lake Nasser. Further east, Sudan seems reluctant to acknowledge Egyptian sovereignty over tribal grazing lands administered by Sudan since 1902.

There are a number of disputes over the precise position of international boundaries in the region. The new Egypt–Israel boundary agreed in 1982 is still the subject of legal and technical disputes at the local level. The classic case of a positional dispute is the boundary between Iran and Iraq which follows the river Shatt al Arab (see Map 48). Several internal boundaries between members of the United Arab Emirates are also in dispute. Territorial arrangements within the federation are complex, reflecting political units derived from tribal relationships (see Fig. 7). There is no reason why boundary disputes should end in conflict. The region abounds with examples of boundary agreements peacefully completed. For example, Saudi Arabia and Jordan exchanged territory in 1965 to give Jordan a longer coastline on the Gulf of Aqaba. In 1966, Saudi Arabia and Kuwait divided the former Neutral Zone. When conflict erupts over boundaries it is usually symptomatic of far deeper political differences.

Table 24. *Types of international land boundary*

	Kilometres	Miles	Per cent
Following physical features	11,938	7,420	35
wadis and rivers	5,792	3,600	
watersheds	4,988	3,100	
edge of plain or plateau	1,158	720	
Following man-made features	1,367	850	4
Following geometric guidelines	19,790	12,300	58
Indeterminate	1,174	730	3
Total	34,269	21,300	100

Source: A. D. Drysdale and G. H. Blake, *The Middle East and North Africa: A Political Geography* (Oxford University Press, New York, 1985).

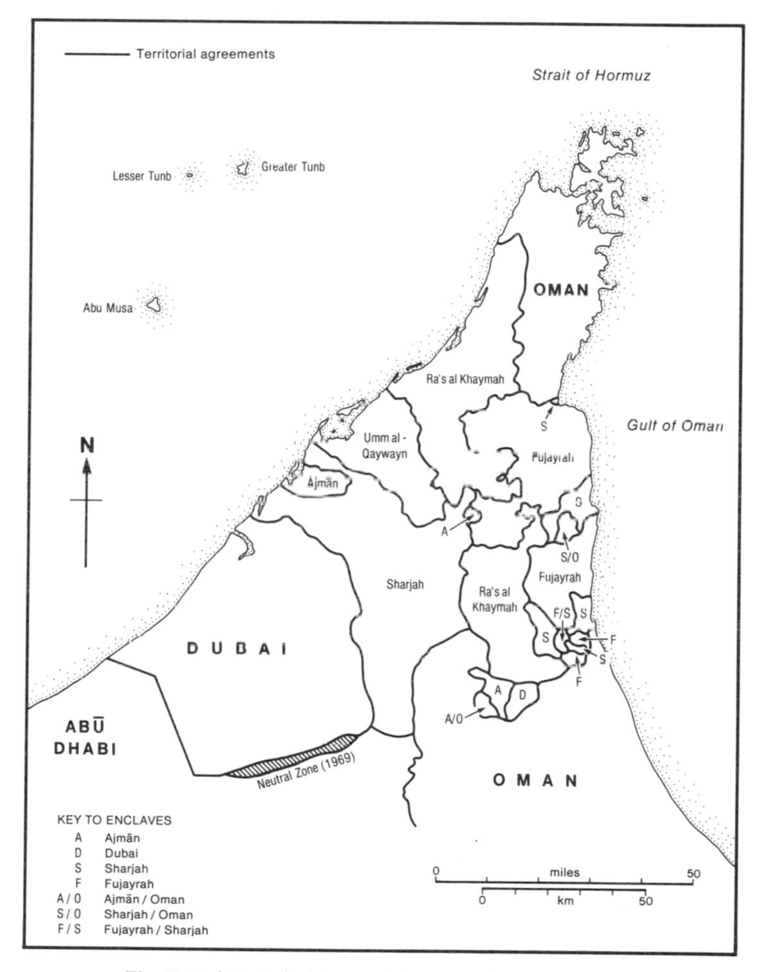

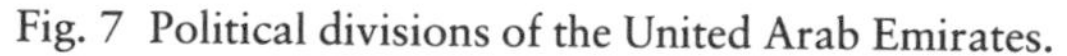
Fig. 7 Political divisions of the United Arab Emirates.

KEY REFERENCES

Brownlie, Ian, *African Boundaries: A Legal and Diplomatic Encyclopaedia* (Hurst and Co., London and University of California Press, Berkeley, for the Royal Institute of International Affairs, 1979).

Drysdale, A. D. and Blake, G. H., *The Middle East and North Africa: a Political Geography* (Oxford University Press, New York, 1985).

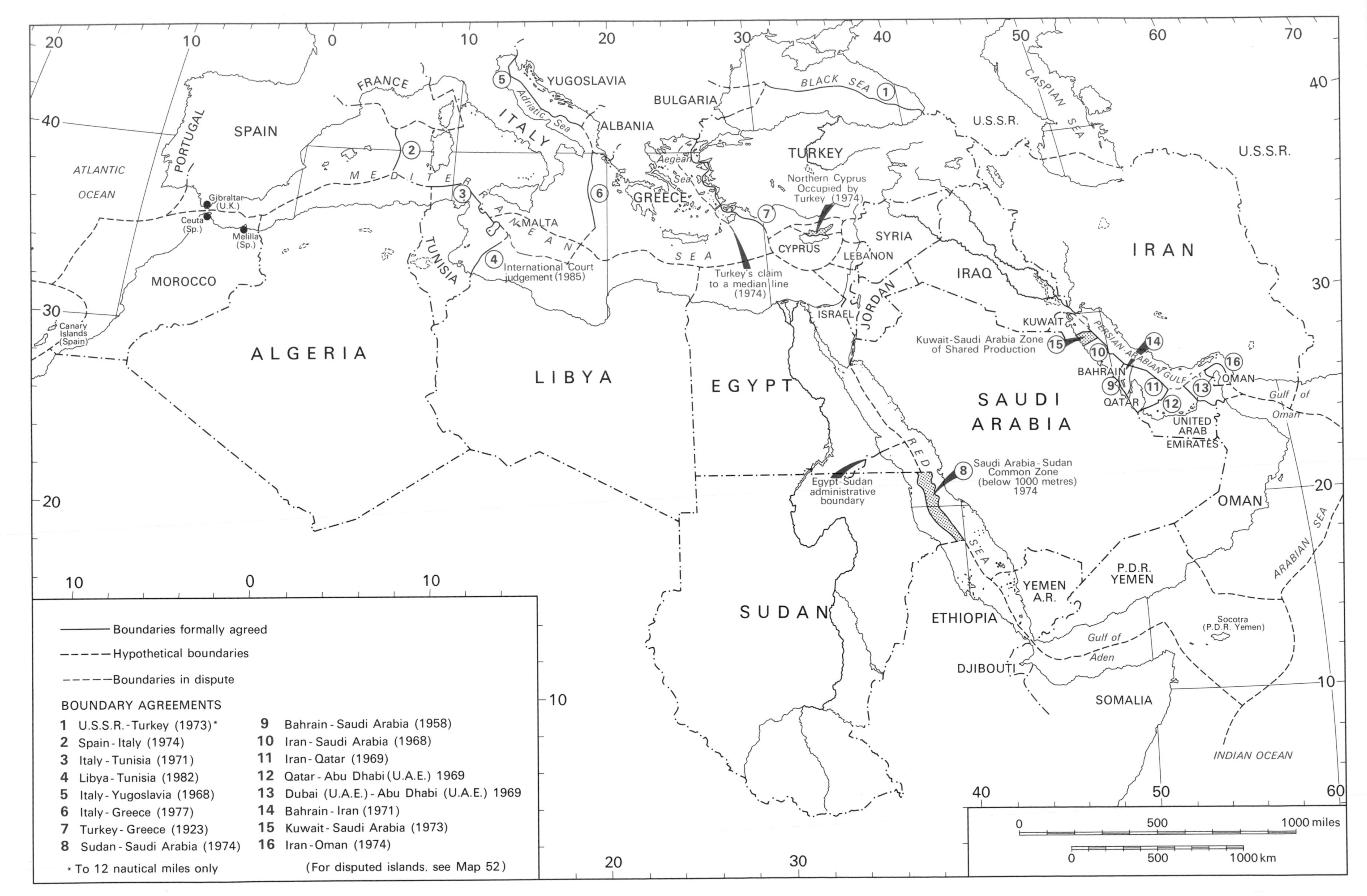

53 Maritime boundaries

Maritime boundaries

Gerald Blake

Most of the land boundaries of the Middle East and North Africa were drawn in the period between 1880 and 1930 (see Maps 16 and 17). Formal maritime boundary delimitation only began seriously in the 1960s and is still far from complete. State jurisdiction offshore is more complicated than on land because differing degrees of state control are recognised, usually embracing internal waters, territorial seas, contiguous zones and continental shelves. Many states also claim exclusive fishing zones. Following the emergence of the United Nations Convention on the Law of the Sea in 1982, after nine years of negotiations, it is necessary to add a sixth type of offshore jurisdiction, the 200 nautical mile exclusive economic zone (E.E.Z.). When the Convention comes into force (which will occur after ratification by 60 states, possibly by 1988), the E.E.Z. may supersede exclusive fishing zones and continental shelf claims, at least as far as the states of the Middle East and North Africa are concerned.

States exercise the same degree of sovereignty over *internal waters* as over the land. Internal waters include bays, estuaries, and waters lying behind any straight baselines which may be drawn along highly indented or island-fringed coasts. These, or in their absence the low-water mark, provide the baseline from which the *territorial sea* is measured. Within territorial water limits, states exercise absolute sovereignty over the air space above, the waterbody, and the seabed, although ships of other states have the right of innocent passage. Middle East and North African states have claimed various breadths of territorial sea in the past, but under the 1982 U.N. Convention it is standardised at 12 nautical miles. Beyond territorial seas, *contiguous zones* are sometimes declared for specific purposes such as the enforcement of sanitary, fiscal and customs regulations. Seven such zones have been declared in the region, mostly to a distance of 18 nautical miles. Coastal states have the right to exploit all the resources of the territorial sea; they are also entitled to the resources of the seabed of their *continental shelves* beyond territorial waters. The presence of oil and natural gas in the seabed of the Gulf provided a strong incentive for the delimitation of some of the earliest maritime boundaries in the region. The proposed E.E.Z. will extend 200 nautical miles offshore from the baseline from which the territorial sea is measured. In the E.E.Z. coastal states have the right to the exclusive exploitation of all living and non-living resources. This significant provision of the 1982 Convention will accelerate maritime boundary delimitation. Only Morocco, Oman and P.D.R. Yemen have the full E.E.Z. entitlement of 200 miles. Elsewhere, the region's narrow seas will have to be partitioned between opposite states.

So far only 17 maritime boundaries between states have been delimited and formally agreed out of at least 70 possible boundaries appertaining to Middle Eastern and North African states. The map shows roughly where these potential boundaries might run, on the basis of median lines. In practice, maritime boundaries do not have to be median lines. Saudi Arabia and Sudan in the Red Sea, for example, opted for a common zone beyond the 1000 metre isobath to facilitate joint exploitation of metal-bearing muds at great depth. Sometimes it may take years for states to devise an equitable alternative to the median line. Thus, in the Mediterranean, Libya rejected the Maltese case for a median line, arguing for an allocation proportional to the length of their opposite coasts. The International Court of Justice tended towards this view in its judgement on the dispute in 1986. The dispute between Greece and Turkey in the Aegean is more complicated. Greece owns nearly all the islands, each with its own territorial water rights. Outside these territorial waters, Turkey claims a share of the seabed up to a median line between the mainlands, as being the 'natural prolongation' of Turkey's land mass. No solution to the dispute is in sight. Where the ownership of the islands is also in contention, offshore boundary drawing is clearly impossible, as in the Gulf between Iraq and Kuwait (see Map 48) and the United Arab Emirates and Iran (see Map 52). Uncertainties over the political future of territories such as Ceuta and Melilla and the United Kingdom bases in Cyprus also hinder seabed allocations.

When the region's seas are finally partitioned between coastal states, their jurisdictional areas will vary considerably, as Table 25 shows. In general, states which enjoy favourable access to the sea will control the greatest areas of seabed. Six of the top eight states in the table have access to more than one sea, giving them enhanced potential for strategic and economic exploitation of offshore areas. Three of the region's vital international waterways (Gibraltar, Bab al Mandeb and Hormuz) are divided by maritime boundaries (see Maps 47 and 51).

Table 25. *Coastal lengths and hypothetical seabed allocations*

	Coastal length		Seabed area (000)	
	Km	Miles	Km^2	$Miles^2$
Oman	1860	1156	561.7	216.9
P.D.R. Yemen	1212	753	550.3	212.5
Libya	1685	1047	338.1	130.5
Morocco	1657	1030	278.1	107.4
Turkey	3557	2211	236.6	91.3
Saudi Arabia	2438	1515	186.2	71.9
Egypt	2422	1505	173.5	67.0
Iran	1833	1139	155.7	60.1
Algeria	1104	686	137.2	53.0
Cyprus	537	334	99.4	38.4
Sudan	718	446	91.6	35.4
Tunisia	1028	639	85.7	33.1
United Arab Emirates	700	435	59.3	22.9
Yemen A.R.	452	281	33.9	13.1
Israel	230	143	23.3	9.0
Qatar	378	235	24.0	9.3
Lebanon	195	121	22.6	8.7
Kuwait	251	156	12.0	4.6
Syria	151	94	10.3	3.9
Bahrain	125	78	5.1	1.9
Jordan	32	20	0.7	0.3
Iraq	58	36	0.7	0.3

Source: A. D. Couper (ed.), *The Times Atlas of the Oceans* (Times Books, London, 1983), p. 227.

KEY REFERENCES

El-Hakim, A. A., *The Middle Eastern States and the Law of the Sea* (Manchester University Press, Manchester, 1979).

Luciani, Giacomo (ed.), *The Mediterranean Region* (Croom Helm, London, 1984).

U.S. Department of State, Office of the Geographer, *National Claims to Maritime Jurisdictions*, Limits in the Seas no. 36, 4th revision (U.S. State Dept, Washington D.C., 1981).

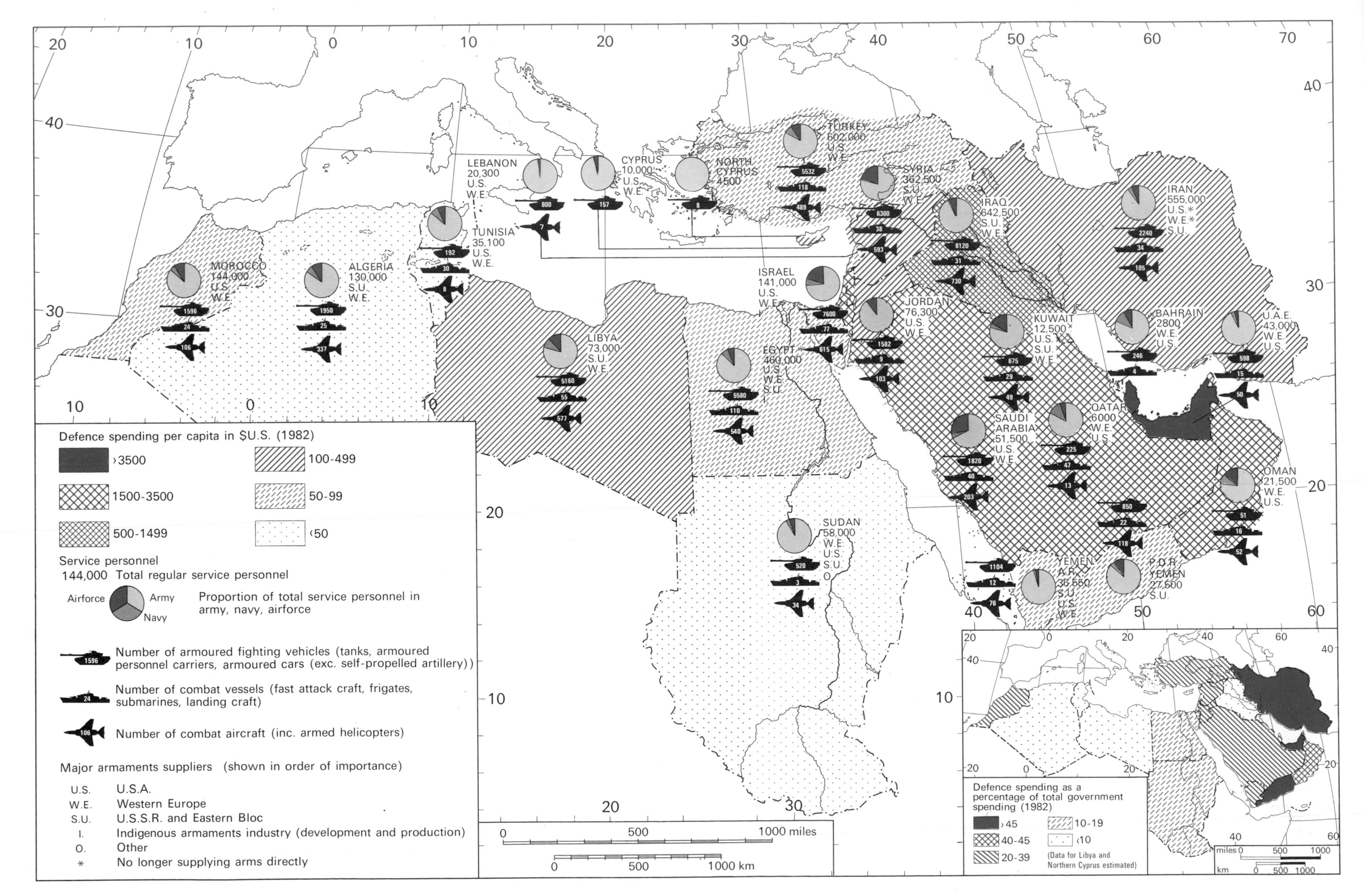
MOROCCO
144,000
U.S.
W.E.
ALGERIA
130,000
S.U.
W.E.
TUNISIA
35,100
U.S.
W.E.
LEBANON
20,300
U.S.
W.E.
CYPRUS
10,000
U.S.
W.E.
NORTH CYPRUS
4500
TURKEY
602,000
U.S.
W.E.
SYRIA
362,500
S.U.
W.E.
IRAQ
642,500
S.U.
W.E.
IRAN
555,000
U.S.*
W.E.*
S.U.
ISRAEL
141,000
U.S.
W.E.
JORDAN
76,300
U.S.
W.E.
KUWAIT
12,500
U.S.
S.U.
W.E.
BAHRAIN
2800
W.E.
U.S.
U.A.E.
43,000
W.E.
U.S.
LIBYA
73,000
S.U.
W.E.
EGYPT
460,000
U.S.
W.E.
S.U.
I.
SAUDI ARABIA
51,500
U.S.
W.E.
QATAR
6000
W.E.
U.S.
OMAN
21,500
W.E.
U.S.
SUDAN
58,000
W.E.
U.S.
S.U.
O.
YEMEN A.R.
36,550
S.U.
U.S.
W.E.
P.D.R. YEMEN
27,500
S.U.
Defence spending per capita in $U.S. (1982)
›3500
1500-3500
500-1499
100-499
50-99
‹50
Service personnel
144,000 Total regular service personnel
Airforce
Army
Navy
Proportion of total service personnel in army, navy, airforce
Number of armoured fighting vehicles (tanks, armoured personnel carriers, armoured cars (exc. self-propelled artillery))
Number of combat vessels (fast attack craft, frigates, submarines, landing craft)
Number of combat aircraft (inc. armed helicopters)
Major armaments suppliers (shown in order of importance)
U.S. U.S.A.
W.E. Western Europe
S.U. U.S.S.R. and Eastern Bloc
I. Indigenous armaments industry (development and production)
O. Other
* No longer supplying arms directly
0 500 1000 miles
0 500 1000 km
Defence spending as a percentage of total government spending (1982)
›45
40-45
20-39
10-19
‹10
(Data for Libya and Northern Cyprus estimated)

54 Defence

Defence

E. W. Anderson

Since 1945, with the exception of Turkey, Tunisia, Qatar and Bahrain, every state in the Middle East and North Africa has been involved in at least one war. The major part of the region constitutes the classical geopolitical fracture zone, long an area of contention between the particular major powers of the day. With the predominance of oil as a world fuel, its importance has increased and it remains the least stable area in the world.

While the conflicts occur elsewhere, there are two major war zones in the core of the region: first, the area of Israel, Syria and Lebanon (Maps 55, 56, 57) and second, the Iran/Iraq frontier (Map 48). Elsewhere, just outside the region, there is continuing conflict in the western Sahara and the Horn of Africa. In each of these cases there has been some superpower influence, most obviously evidenced by the source of weapons. On the other hand, it is clear that over the recent past, many states, notably Egypt, Syria, Yemen A.R., Iran, Iraq, Libya, Sudan and Algeria, have received weapons from both East and West. Recently Kuwait joined the list as the first Gulf Co-operation Council state to trade with both sides. Within the region, only P.D.R. Yemen is armed solely by the Soviet Union.

Despite the obvious competition for influence in the region, only Turkey, as a member of N.A.T.O., belongs to one of the two major power blocs. However, there are key bases within the region. The United States has bases in Oman and Bahrain within the region and in Somalia just outside it. The United Kingdom retains bases in Oman and Cyprus, and Spain in its Moroccan enclaves. The only Soviet bases are in P.D.R. Yemen and, adjacent to the region, in Ethiopia. In terms of broader alignments, Libya, Syria and P.D.R. Yemen have ties with the East. The remaining states, with the exception of Algeria and Iran which are neutral, are pro-West.

In such an unstable environment, the temptation for outside interference must always be present, and to the West the major destabilising force would seem to be the Soviet Union. Its move into Afghanistan gave it a forward position nearer the centre of Gulf activity. The immediate cause was probably instability in what had been effectively a client state along its own borders. However, given its already favourable position of influence throughout Ethiopia and P.D.R. Yemen, the advance has also been interpreted as the early stage of a pincer strategy for cutting off Gulf oil and also for erecting a boundary to any westward diffusion of Chinese interests. Should Afghanistan be sufficiently subdued to provide a stable base, the military advantage gained for the Soviet Union would be comparatively small, but there would be a platform for subversion in the region. The fact remains that the Soviet Union is not as well represented in the mid-1980s as it was in the 1960s when its influence was well entrenched in both Egypt and Iraq, giving the core of the region an Eastern orientation.

The other main component in the argument is the recent short-fall revealed in Soviet oil production which has been interpreted by some as the start of a continuing decline. If this were so, access to Gulf oil might soon become crucial. On the other hand, the present circumstances may be attributable more to bad management, a hostile environment and lack of sophisticated technology, rather than terminal problems. There is obvious potential for further oil exploration in the Soviet Union itself and, as holder of by far the world's largest published proven reserves of natural gas, substitution in the energy base for certain industries is a definite option. Thus current oil difficulties are probably temporary, but towards the end of the century the Soviet Union could become increasingly reliant upon the Persian/Arabian Gulf. At present there seems little chance of direct Soviet intervention within the region, particularly as this would result in a confrontation with the stated vital interests of the United States. Therefore current Soviet activity may best be interpreted as capitalising on natural instability which, sadly, appears characteristic of the region.

To the inhabitants of the region, the United States, through its support for Israel and its potential for military action, also appears as something of a destabilising factor. Furthermore, the Central Command, with its rapid deployment force, poses a dilemma which has already divided the Arab states so that only Oman, which provides bases, and Egypt have staged collaborative simulation exercises.

A graver threat to the region is posed by the inter-state disputes which have led to so many boundary conflicts. Notable among these are the dormant Iranian claims to Bahrain and also the Tumb Islands, the dispute between Qatar and Bahrain over Huwar Island and the continuing confrontation over islands in the approach to the port of Umm Qasr between Iraq and Kuwait. Border conflicts also occurred in Oman, at Buraimi in the north and in Dhofar to the south. With incomplete boundary delimitation, both on land and at sea, such conflict is likely to continue.

Since 1945 there have been at least 21 major border wars. Many such wars, particularly those around Israel, Lebanon and Libya and the border shared between Iran and Iraq, remain scenes of hostility. Where other factors such as major religious, political or racial differences are superimposed, the problems seem even more intractable. In a region where oil, gas and water are so vital, the provision of clear-cut boundaries is obviously critical.

Any of the disputes might result in inter-state conflict, bringing interruptions to oil flow through damage to the actual installations or interference with the regional infrastructure. As increasingly more oil traverses the region by pipeline, so the vulnerability increases (Map 46). The other major element is, of course, the sea, and both Hormuz and Bab al Mandeb are not only narrow and congested, but are bounded by mutually hostile riparian states (Map 47).

Despite the foregoing analysis, the major threats to security would seem to be internal. The political structures, particularly of the Gulf region, seem at least in the long term unstable. Variations in religion and life style, together with differences resulting from the rapid rate of modernisation and the necessity for expatriate labour, have frequently produced tensions. Also in many states there are significant minorities, most notably the Kurds who inhabit parts of Turkey, Syria, Iraq and Iran. Thus military strength has also been developed for counter-insurgency operations.

In the early 1980s within the region, only Turkey and Yemen A.R. were under military rule, but since 1945, nine other countries have had military governments. Every country in the region has witnessed some level of civil war, except the U.A.E., Qatar and Bahrain, while major uprisings have occurred in Morocco, Cyprus, Syria, Iraq and Iran. In a move towards greater self-sufficiency, the region has become a producer of arms. Only eight countries have no arms production industry, while Turkey, Egypt and especially Israel both develop and produce a wide range of armaments. Egypt, Israel and Saudi Arabia are providers of extensive military advice and training, while at least four other states are involved on a lesser scale.

Furthermore, the economies themselves are vulnerable, not only as a result of unfulfilled expectations, but also because basic requirements from indigenous sources are scarce. As essentially non-agricultural states, the smaller Gulf countries rely upon producers such as the Lebanon for their basic food supplies. Even water is increasingly at risk as a result of virtually uncontrolled pumping for many years. Ground water supplies are everywhere greatly reduced and there is an increasing reliance upon desalination plants, particularly soft targets in any period of conflict.

KEY REFERENCES

Cordesman, A. H., *The Gulf and the Search for Strategic Stability* (Westview Press, Boulder, Colorado, 1984).
Kidron, M. and Smith, D. *The War Atlas* (Pan Books, London, 1983).

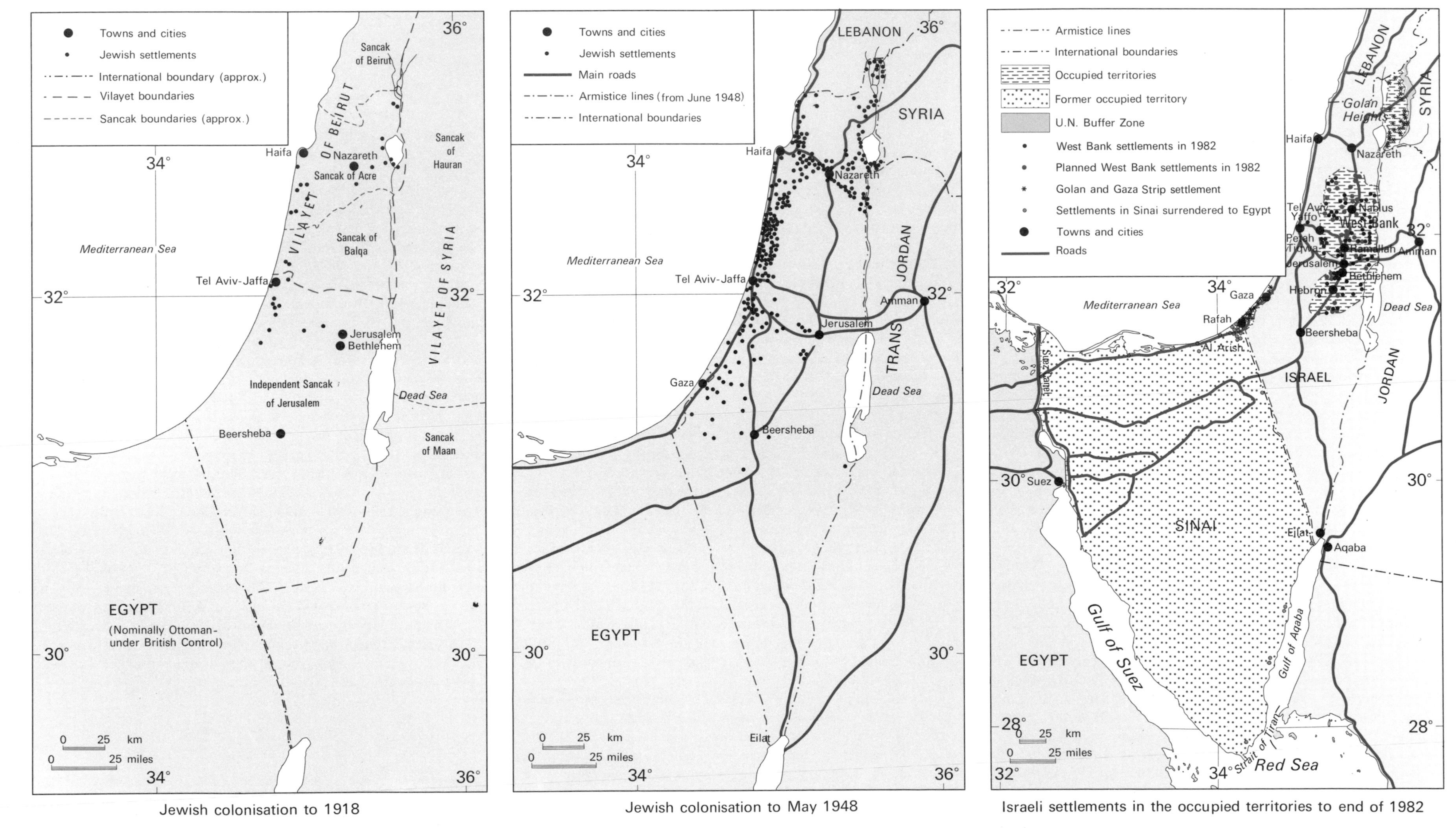

Jewish colonisation to 1918

Jewish colonisation to May 1948

Israeli settlements in the occupied territories to end of 1982

55 Palestine

Palestine and Israel

Gerald Blake

Jewish colonisation

In the late nineteenth century Palestine was still part of an extensive but declining Ottoman Empire. For administration it was divided into three northern Sancaks (Beirut, Acre and Balqa) which were part of the larger Vilayet of Beirut, with the independent Sancak of Jerusalem in the south (Map 55A). East of the Jordan–Dead Sea rift valley was the Vilayet of Syria, subdivided into the Sancaks of Hauran and Maan. In 1906 the British in Egypt and the Ottomans delimited a boundary which left the Sinai peninsula as part of Egypt.

Palestine was a small and impoverished territory with approximately 500,000 Arab inhabitants in the 1880s. Jewish colonisation began in 1882, spurred on by persecutions in Russia and Eastern Europe, and by the Zionist movement among European Jews. By 1918 there were 644,000 Arabs (92%) and 56,000 Jews (8%) in Palestine. Many Jews settled in the towns, but more than 40 Jewish rural settlements were established, chiefly in the lowlands, on approximately 2% of the land. Map 55A shows both the privately owned villages (or *moshavot*) and the earliest collective farms (or *kibbutzim*). The latter were to become ideal for colonisation in dangerous locations and difficult environments. World War I created conditions for large-scale Jewish immigration into Palestine. The Ottomans were driven out by British and Arab forces, and in 1920 Britain was given a League of Nations Mandate to govern Palestine. The Mandate required Britain to encourage the establishment of a Jewish national home in Palestine in accordance with promises incorporated in the Balfour Declaration of November 1917. The Palestine Arabs deeply resented this policy and never ceased to register their opposition, particularly in the 1930s when Jewish refugees arrived from Germany in ever-increasing numbers (66,000 in 1935, for example). British attempts to restrict Jewish immigration and land transfer to Jews were largely ineffective and by May 1948 almost 6% of the land of Palestine was owned by Jews, and a higher proportion of the cultivable land. Map 55B shows Palestine's boundaries and the pattern of Jewish colonisation during the British mandate. Boundaries with Lebanon were agreed in an Anglo-French Convention in 1920 and delimited on the ground in 1923. The boundary with Transjordan was fixed in 1922. The 1906 boundary with Egypt remained unchanged. Palestine thus defined covered 25,900 km^2. It was not co-terminous with biblical Palestine nor with any former Ottoman administrative districts.

Jewish colonisation in the Mandate period was heavily concentrated in the lowlands, avoiding the populous uplands of Galilee, Judaea and Samaria and the inhospitable Negev desert in the south. After the end of the British Mandate in May 1948, fierce fighting between Jews and Arabs left Israel in control of 77% of former Palestine, instead of the 56% allotted to the Jews under the United Nations partition plan of November 1947 (Map 56A). Some 600,000 Arab refugees fled from Israel during hostilities, leaving about 160,000 Arabs inside the new state of Israel. By the end of 1948, Israel's population had reached 915,000, 83% Jews and 17% Arabs. Since 1948, the settlement pattern and population structure of Israel have been transformed. With the influx of 1.8 million Jews (38% before 1952) rural colonisation and urban growth accelerated. About one-third of the immigrants were dark-skinned 'Oriental' Jews from the Middle East and North Africa. Over 500 rural settlements have been founded, particularly *moshavim* and *kibbutzim*, many of them as part of extensive rural settlement projects in areas sparsely populated by Jews like the Negev and Galilee. Some 40 new towns were also built as regional service centres and development foci outside the major cities of Tel Aviv, Jerusalem and Haifa. Israel's Jewish population in 1986 was nearly 90% urban, one of the world's highest ratios. Jewish immigration has fallen markedly in recent years, which is politically significant because Israel's Arab population has a very high rate of natural increase (about 3.4% per annum).

The Arab–Israel war of June 1967 left Israel in possession of Sinai and the Gaza Strip, the Golan Heights and the West Bank (Map 56B). All three areas were subsequently colonised by Israel though for different motives. Some 22 settlements totalling perhaps 4000 inhabitants were established in Sinai, mostly associated with tourism (Gulf of Aqaba) and agriculture (Rafah region). These were evacuated when Sinai was returned to Egypt (Map 56C), though not without bitter protests from their inhabitants. Several small Jewish villages remain in the densely populated Gaza Strip where 500,000 Arabs live. The Golan Heights were almost completely evacuated by their 100,000 Syrian inhabitants in 1967. Settlements there provide northern Israel with a first line of defence against Syrian invasion, although the total Jewish population is small. Parts of eastern Golan might be negotiable in the event of peace with Syria, but the western area overlooking Israel is unlikely to be surrendered.

Israel's emotional and geographical ties with the West Bank make it the most prized of the Arab lands occupied in 1967. The region has a substantial Arab population (1984 estimate 850,000) and clearly constitutes a potential heartland for any future Palestinian state, with Jerusalem as its natural focus. Until 1977, Israeli settlement strategy under Labour governments followed the Allon Plan designed to establish a line of strategic settlements, mostly *kibbutzim* and *moshavim*, along the Jordan valley, but avoiding the populous interior mountain area (Map 55C). Pressure mounted, however, for more widespread settlement of Judea and Samaria on historic and religious grounds, and from 1974 the Gush Emunim movement began founding illegal settlements. The Likud government, elected in 1977, broadly supported this policy and funded the establishment of a large number of Jewish settlements. In 1980, the lifting of restrictions on private land purchases by Jews greatly accelerated West Bank settlement so that by 1985 (Map 55C) there were over 90 settlements at various stages of planning and development. The West Bank Jewish population rose from 12,000 in 1980 to 45,000 in 1985. Long-term plans for the region envisage dozens more villages and half a dozen towns accommodating one million Jews. According to the influential Drobless plan of 1978, settlement effort is to be concentrated into a number of consolidated blocks (Map 55C). The eastern third, and areas adjacent to Israel, are marked out for special attention. Most Jewish settlements established since 1977 are of the *yishuv kehillati* ('community settlement') type, based upon light industrial enterprises and commuting to urban centres in Israel, with only minimal agricultural activity.

Israeli penetration of the West Bank is now very extensive, and possibly irreversible. Although less than 3% of the land is earmarked for immediate settlement, Israel controls more than half the land on the West Bank. Much was formerly Jordanian government land, and large areas have been requisitioned for military and public purposes, often in spite of local protests. There is an extensive network of new roads, and electricity and water services reach Jewish settlements. By contrast, Arab settlements have been poorly treated and several village water supplies have been adversely affected by Jewish settlement activity. Although Jewish numbers are relatively small (compared with the annual increase of Arab population, for example), Israel's capital investment in the West Bank is substantial. This is especially evident in the large suburbs around Jerusalem, which the Israelis clearly do not intend to hand back to the Arabs.

Territorial expansion of Israel

The British government promised to facilitate the establishment of a Jewish national home in Palestine in the Balfour Declaration of 2 November 1917, while recognising the civil and religious rights of the non-Jewish population. These incompatible objectives created conflict which Britain was unable to resolve. In 1937, Lord Peel's Royal Commission recommended partition; in 1938, the Woodhead Commission proposed an alternative scheme (Fig. 8). Neither scheme was workable, not least because the populations were incompletely

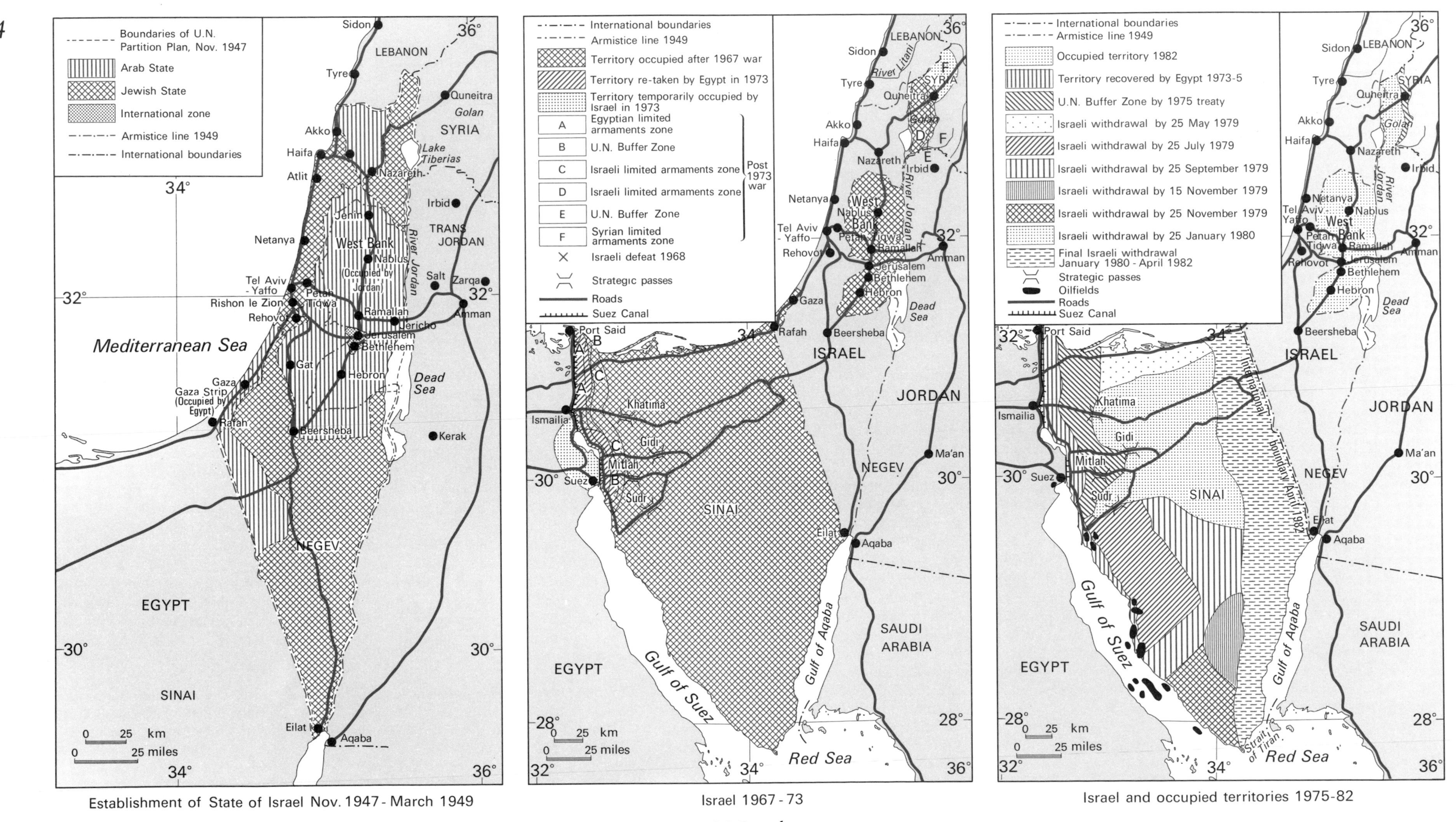

Establishment of State of Israel Nov. 1947 - March 1949

Israel 1967 - 73

Israel and occupied territories 1975-82

56 Israel

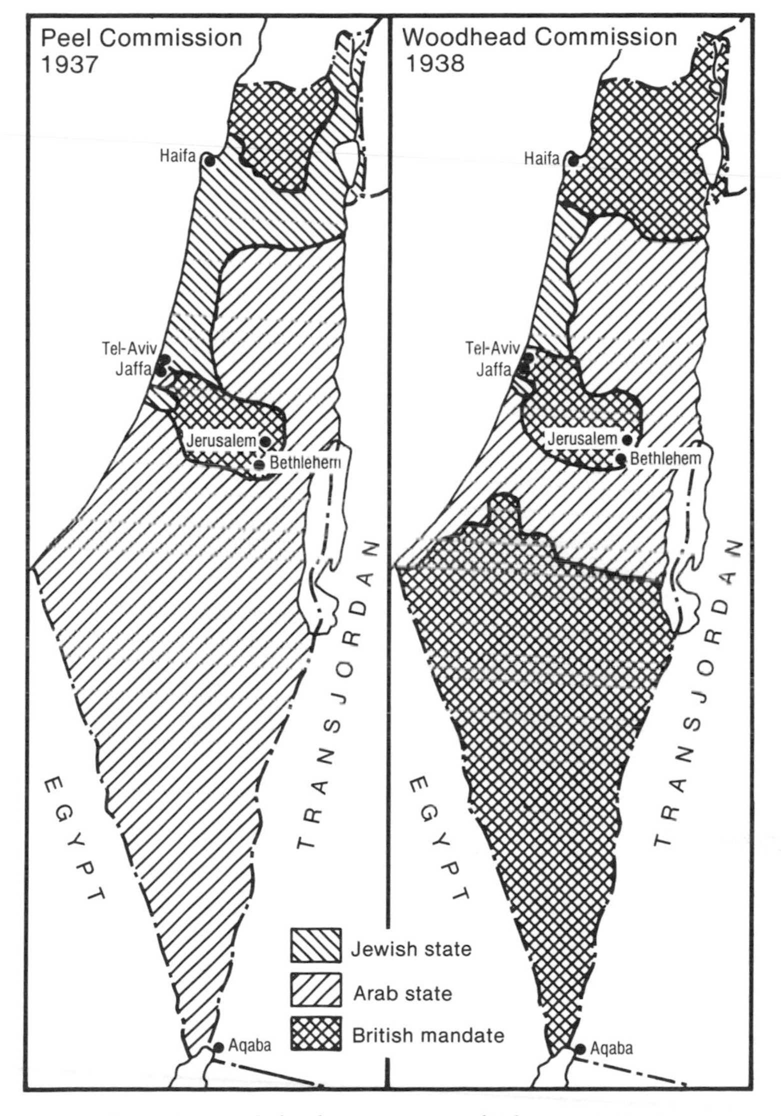

Fig. 8 Proposals for the partitioning of Palestine, 1937–8.

segregated. Although Britain restricted Jewish immigration and land transfers thereafter, there was no partition. Zionist colonisation intensified, however, notably in the Negev. Thus the United Nations Partition proposal of 29 November 1947 gave the Jewish state more territory than did the British proposals. Although the Jewish state was fragmented, Yaffo remained an Arab enclave and Jerusalem was internationalised (Map 56A); the proposal was acceptable to the Jews. Israel proclaimed independence on 14 May 1948. Several months of fighting followed between Israel and neighbouring Arab states, ending with armistice agreement in 1949. The area occupied by Israel (20,700 km^2) was about one-fifth larger than under the U.N. plan, and was unchanged until June 1967.

From an Israeli viewpoint the 1949–67 boundaries were unsatisfactory. State security was threatened by narrow corridors in the centre and south. Infiltration from the Gaza Strip and elsewhere was commonplace. Jerusalem was a divided city, while in the north Syrian forces could shell Israeli settlements at will. In May 1967, Israel became convinced that Arab forces were mobilising for war, and on 5 June launched a pre-emptive air strike as a prelude to the conquest of Sinai and Gaza (61,558 km^2), the West Bank (5900 km^2) and Golan Heights (1250 km^2), increasing Israeli-controlled territory threefold, while actually reducing the land boundaries of the state. The strategic and economic gains were considerable, including acquisition of the Sinai oilfields and a large reservoir of Arab labour. Many states paid a heavy price, however, because of the closure of the Suez Canal until 1976 (Maps 46 and 50).

Between June 1967 and 1970, Israeli forces along the Suez Canal and Jordan Valley sustained heavy casualties in a war of attrition involving artillery and cross-border raids. On 6 October 1973, Syria and Egypt launched a co-ordinated invasion of Golan and Sinai with devastating effectiveness. In counter-attacks, Israeli forces penetrated far into Syria and crossed the Suez Canal (Map 56B). An Israel–Syria disengagement agreement (31 May 1974) finally brought about Israeli withdrawal, and the introduction of a U.N. Buffer Zone and zones of limited forces. A ceasefire was arranged in Sinai on 24 October 1973 and a first disengagement agreement was reached in January 1974, again involving a U.N. zone and zones of limited forces (Map 56B). In September 1975 a second Sinai agreement extended the U.N. zone to include the crucial Gidi and Mitla passes (Map 56C), and extended Egypt's front line eastwards. As part of the Israel–Egypt agreement signed at Camp David in September 1978, Israel was to withdraw from Sinai in phases. The final phase was completed in April 1982. Egypt agreed to allow Israel shipping through the Suez Canal and access to the Gulf of Aqaba. The international boundary between Egypt and Israel was formally agreed, with a few minor difficulties (as at Taba on the Gulf of Aqaba coast). The future of the other occupied territories remains obscure, and somewhat gloomy.

KEY REFERENCES

Gilbert, M., *The Arab–Israel Conflict: Its History in Maps*, 3rd edn (Weidenfeld and Nicolson, London, 1979).

Harris, W. W., *Taking Root: Israeli Settlement in the West Bank, the Golan and Gaza–Sinai, 1967–80* (Research Studies Press, Chichester, 1981).

Newman, D. (ed.), *The Impact of Gush Emunim: Politics and Settlement in the West Bank* (Croom Helm, London, 1985).

Orni, E. and Efrat, E., *Geography of Israel*, 3rd edn (Israel Universities Press, Jerusalem, 1971).

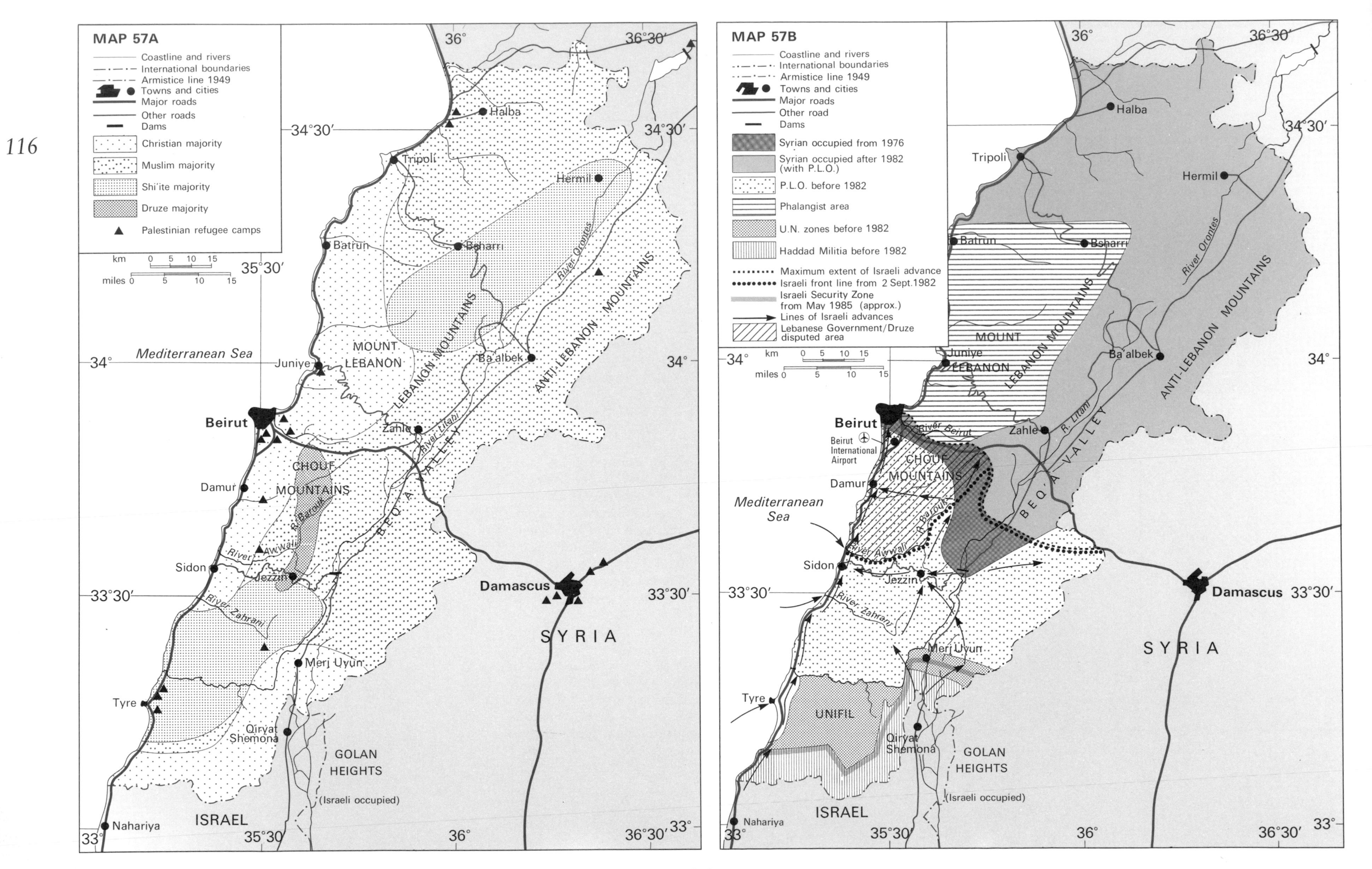
MAP 57A
Coastline and rivers
International boundaries
Armistice line 1949
Towns and cities
Major roads
Other roads
Dams
Christian majority
Muslim majority
Shi'ite majority
Druze majority
Palestinian refugee camps
km 0 5 10 15
miles 0 5 10 15
Mediterranean Sea
Halba
Tripoli
Hermil
Batrun
Bsharri
MOUNT LEBANON
LEBANON MOUNTAINS
ANTI-LEBANON MOUNTAINS
River Orontes
Juniye
Ba'albek
Beirut
Zahle
River Litani
BEQ'A VALLEY
CHOUF MOUNTAINS
Damur
R. Barouk
River Awwali
Sidon
Jezzin
Damascus
SYRIA
River Zahrani
Merj Uyun
Tyre
Qiryat Shemona
GOLAN HEIGHTS
(Israeli occupied)
Nahariya
ISRAEL
MAP 57B
Coastline and rivers
International boundaries
Armistice line 1949
Towns and cities
Major roads
Other road
Dams
Syrian occupied from 1976
Syrian occupied after 1982 (with P.L.O.)
P.L.O. before 1982
Phalangist area
U.N. zones before 1982
Haddad Militia before 1982
Maximum extent of Israeli advance
Israeli front line from 2 Sept.1982
Israeli Security Zone from May 1985 (approx.)
Lines of Israeli advances
Lebanese Government/Druze disputed area
River Beirut
Beirut International Airport
R. Litani
UNIFIL

57 Lebanon

MAP 57

Lebanon

Gerald Blake

The geography of the tiny state of Lebanon (10,400 km^2) helps explain its turbulent political history and its relative prosperity. Lebanon controls routes from the interior to the coast, and from north to south along the narrow coastal plain. Two well-watered mountain ranges (the Lebanon and Anti-Lebanon Mountains) are separated by the fertile Beq'a valley. There are numerous secure sites which attracted various persecuted minorities to Lebanon over the centuries and helped ancient Christian communities to survive. Thus Lebanon today displays extraordinary religious diversity, with 17 confessional groups being recognised. In addition to these, there are large Palestinian refugee communities (perhaps 350,000 in 1983). The precise size of the confessional groups and communities is uncertain. No census has been taken since 1932 and migration and high mortality have complicated estimates in the past decade.

Lebanon was carved out of Syria by the French Mandatory power in 1920. In the nineteenth century, the Ottoman administration had given semi-autonomous status to the Mutasarrifiyah of Mount Lebanon, with a Christian governor. France expanded Mount Lebanon in an attempt to create a more viable Christian state with secure frontiers. The enlarged Lebanon included considerable numbers of Shi'ite and Sunni Muslims. Lebanon declared independence in 1941. In 1943, an elaborate scheme of power sharing was devised between Lebanon's rival confessional groups but, below the surface, traditional communal and kinship loyalties were far stronger than loyalty to the state. As Lebanon prospered after World War II, socio-economic disparities between Christians and Muslims grew, and there was growing resentment of the political power of the Maronites, particularly as the demographic balance had clearly shifted in favour of the Muslims (Table 26). Map 57A shows Lebanon on the eve of the civil war, which finally broke out in 1975, after an earlier bout of fighting in 1958. Christians were largely concentrated in the most favoured coastal and Mount Lebanon areas, and Muslims in the poorer peripheral regions. About 30% of the national population lived in Beirut, including large camps of Palestine refugees. Increasing numbers of poor Shi'ite and Sunni rural migrants settled in the suburbs of the city.

The civil war had cataclysmic consequences for Lebanon. Central government collapsed, the once buoyant economy was ruined, the Lebanese army disintegrated, and in the first year of fighting 50,000 people died and over 100,000 were injured. The Syrian army intervened at the request of the President of Lebanon in 1976, and remains firmly in control of northern and eastern Lebanon (Map 57B). Other spatial effects of the war are also shown on the map. A Maronite mini-state appears north of Beirut, dominated by Phalangist militias who emerged as the dominant force among the Maronites. In the south, a large tract of territory was effectively controlled by the Palestine Liberation Organisation (P.L.O.).

Israel invaded southern Lebanon in March 1978 to tackle P.L.O. strongholds, but withdrew under strong pressure from the United States. A United Nations Interim Force in Lebanon (U.N.I.F.I.L.) comprising 7000 men was deployed to preserve Lebanon's territorial integrity. Israel also sponsored Christian militias in the southern borderlands in an attempt to create a 'security zone'. A second Israeli invasion was launched in June 1982, designed to destroy the P.L.O. and ensure the establishment of a strong Phalangist government with whom a lasting Israel–Lebanon peace might be negotiated. The P.L.O. was forced to withdraw from Beirut in August 1982 and its power in Lebanon has been broken, at least for the time being. To this extent Israeli objectives were achieved, but the price was very high. The deaths of 12,000 Lebanese and 700 Israelis brought widespread condemnation, and bitter dissension inside Israel. Syrian influence in Lebanon has been strengthened, and that of Israel's Maronite allies much weakened. The Druzes succeeded in clearing the Chouf Mountains of all remaining Christian influence during 1983 (Map 57B). Israel's invasion also revealed the political and military power of the Shi'ites, who now form the largest single group in Lebanon.

Lebanon's political dilemmas will be difficult to resolve. Partition would not appeal to most Lebanese. A confederation of cantons based on traditional community heartlands might work, but there would be difficulties in Beirut, where almost one in three Lebanese now live.

Table 26. *Lebanon: population by community*

Religion	No.	Percentage 1983	Percentage 1932
Christians			
Maronites	900,000	25.1	29.0
Greek Orthodox	250,000	7.0	10.0
Greek Catholic	150,000	4.2	6.0
Armenians	175,000	4.9	4.0
Others	50,000	1.4	3.0
Total Christians	1,525,000	42.6	52.0
Muslims			
Shi'ites	1,100,000	30.8	19.0
Sunnis	750,000	21.0	22.0
Druzes	200,000	5.6	7.0
Total Muslims	2,050,000	57.4	48.0

Source: D. McDowall, *Lebanon: A Conflict of Minorities*, Minority Rights Group Report no. 61 (M.R.G., London, 1983).

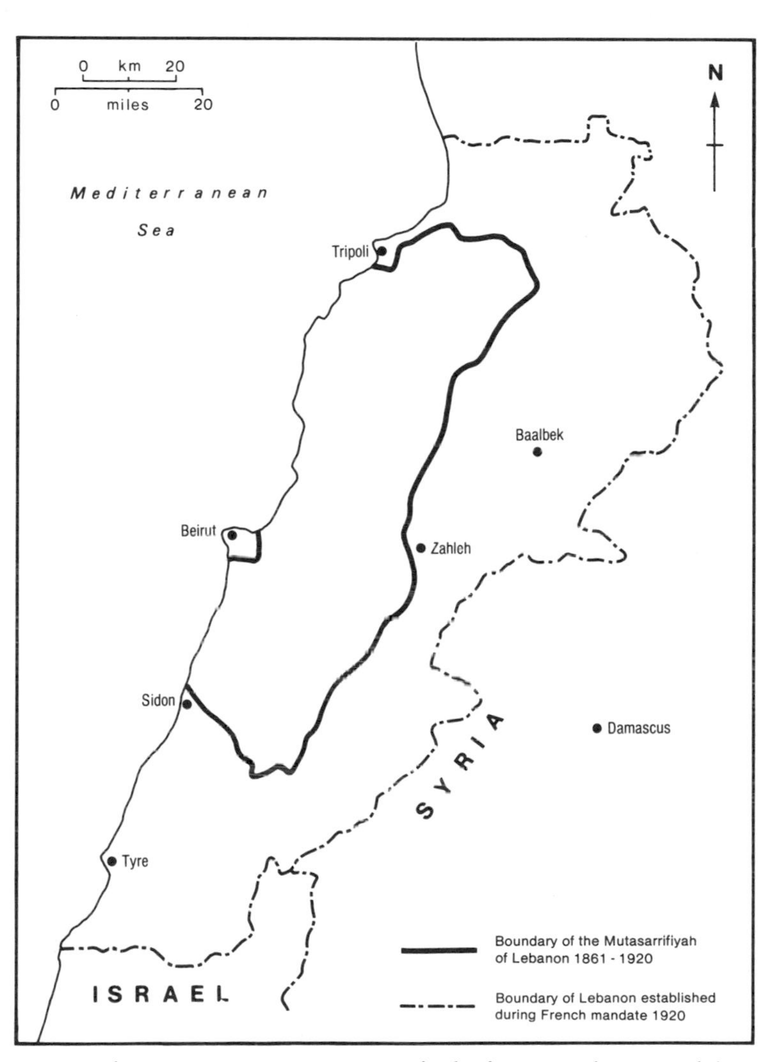

Fig. 9 The semi-autonomous Mutasarrifiyah of Mount Lebanon, and the boundaries of modern Lebanon.

KEY REFERENCES

Gilmour, D., *Lebanon: The Fractured Country* (Sphere Books, London, 1983).

McDowell, D., *Lebanon: A Conflict of Minorities*, Minority Rights Group Report no. 61 (M.R.G., London, 1983).

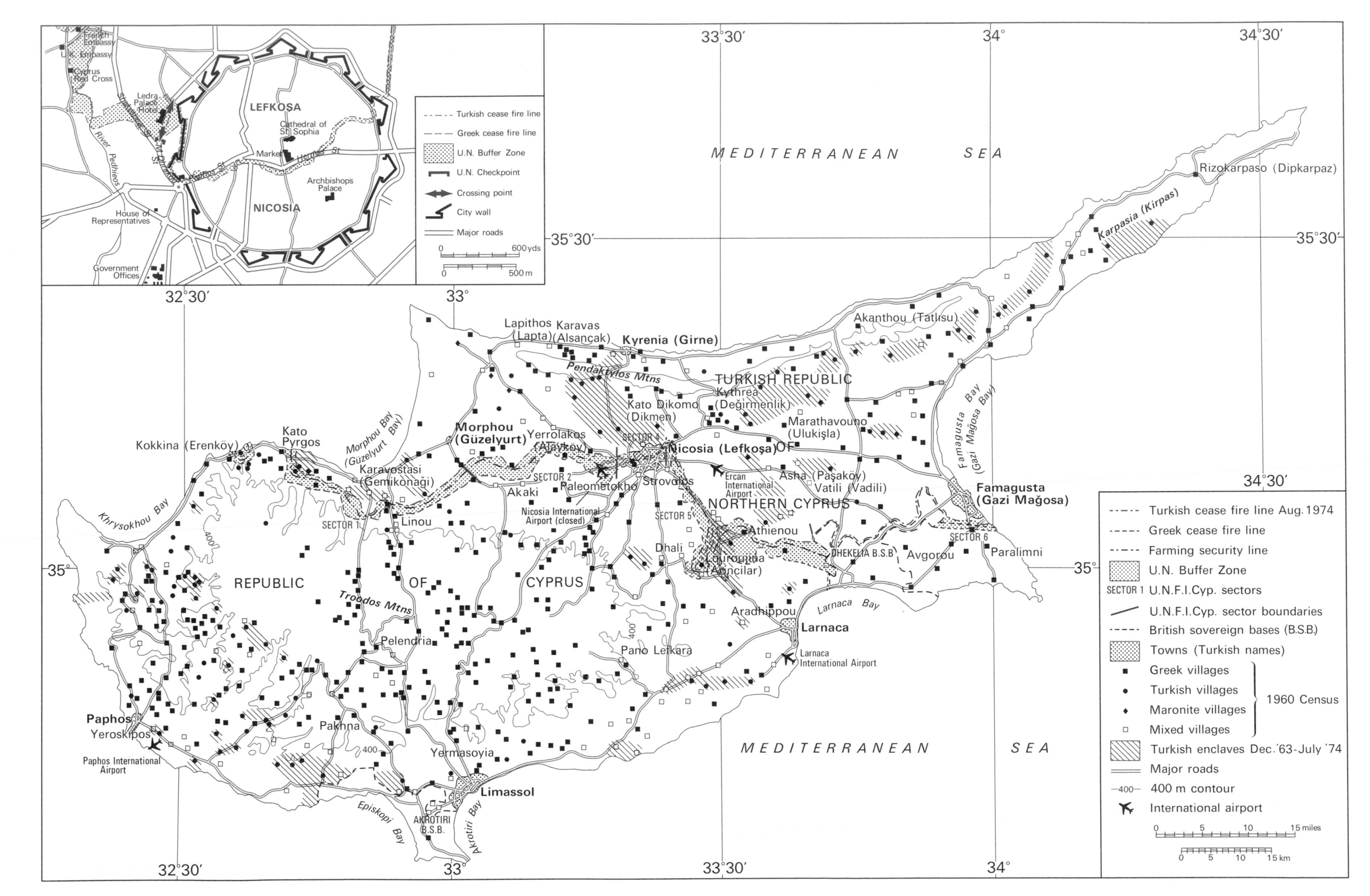

MEDITERRANEAN SEA
TURKISH REPUBLIC OF NORTHERN CYPRUS
REPUBLIC OF CYPRUS
Nicosia (Lefkoşa)
Kyrenia (Girne)
Famagusta (Gazi Mağosa)
Larnaca
Limassol
Paphos
Morphou (Güzelyurt)
Troodos Mtns
Pendaktylos Mtns
LEFKOSA
NICOSIA
Turkish cease fire line Aug. 1974
Greek cease fire line
Farming security line
U.N. Buffer Zone
SECTOR 1 U.N.F.I.Cyp. sectors
U.N.F.I.Cyp. sector boundaries
British sovereign bases (B.S.B.)
Towns (Turkish names)
Greek villages
Turkish villages
Maronite villages
Mixed villages
1960 Census
Turkish enclaves Dec.'63-July '74
Major roads
400 m contour
International airport

58 Cyprus

Cyprus

Carl Grundy-Warr

Twice since independence from British colonial rule in August 1960, the political map of Cyprus has altered dramatically, resulting in bloodshed and population upheavals.

Serious inter-communal clashes began during the late 1950s when Greek Cypriot agitation for self-determination and *Enosis* (Union with Greece), led by E.O.K.A. (National Organisation of Freedom Fighters), was at its peak. *Enosis* was anathema to the Turkish Cypriots who did not want to become a minority ruled over by Athens. Owing to the island's strategic position in the eastern Mediterranean close to Turkey's shores, the Cyprus problem was causing a rift within N.A.T.O. between Greece and Turkey. Eventually tripartite talks between Britain, Greece and Turkey led to the London–Zurich Agreements of 1959, which laid the basic framework of the independence constitution. Britain retained a foothold in the region under the Treaty of Establishment, which provided for two British sovereign base areas, Akrotiri and Dhekelia, totalling 256 km^2 of Cypriot territory.

At independence, Cyprus had a population of 650,000, Greeks out-numbering Turks four to one. The island had some 110 mixed villages, and distinct ethnic quarters in the main towns and in several villages. There was peaceful co-existence for the Republic's first three years until the fragile Constitution designed to preserve the island's bicommunal character broke down. On 30 November 1983, President Makarios wrote to Vice-President Kücük proposing 13 amendments to the constitution, which if implemented would have transformed the Republic into a state ruled by majority principles, demoting the status of Turkish Cypriots from 'co-founders' of the Republic to that of a minority, and opening the gates to *Enosis*. Soon after this political cleavage, inter-communal violence flared up in Nicosia. By 28 December 1963 a Green Line had been drawn across the capital to act as a temporary *cordon sanitaire* between the communities, but it soon became a symbol of ethnic segregation and a division of international geopolitical significance.

As violence spread to other towns and villages, approximately 25,000 Turkish Cypriots evacuated isolated and mixed villages for Turkish Cypriot enclaves protected by their own paramilitaries. Between December 1963 and August 1964 a patchwork partition had developed. The Turkish Cypriot leadership, having withdrawn from the official Government of the Republic, effectively controlled a rudimentary, fragmented 'state within a state' comprising tiny enclaves scattered throughout the island. When the United Nations peace-keeping force (U.N.F.I.Cyp.) took over from a beleaguered British force on 27 March 1964, the island was criss-crossed by barbed-wire fences, road blocks and other fortifications, obstacles to free movement and normal life. U.N.F.I.Cyp.'s initial task was to interpose troops between the Greek National Guard and Turkish Cypriot fighters. Despite numerous U.N.F.I.Cyp. efforts to 'normalise' conditions, the two communities drifted further apart and the physical dividing lines remained intact for a decade. The introverted Turkish enclaves were too small to be economically viable and their isolation from the rest of the Cypriot economy reinforced the growing economic dichotomy between the two communities.

Meanwhile the political situation deteriorated as tensions rose between Makarios and the Greek military Junta. The Archbishop narrowly escaped death in an Athens-inspired *coup d'état* against him on 15 July 1974. The conspirators installed Nicos Sampson, a former E.O.K.A. anti-Turk terrorist, as head of the new 'regime of national salvation'. On 20 July, Turkey exercised her right to intervene in Cyprus according to Article 4 of the Treaty of Guarantee by launching a military invasion. A second Turkish military advance began on 13 August after the collapse of negotiations at Geneva between the Guarantor Powers and Cypriot representatives. By the time an island-wide cease fire was called on 18 August the Turks had succeeded in occupying 37% of Cyprus. One immediate result of this arbitrary partition was a massive population exchange. Between July 1974 and December 1975, about 185,000 Greek refugees moved south and 45,000 Turkish Cypriots moved north. Northern Cyprus was named the 'Turkish Federated State of Cyprus' on 13 February 1975, a *de facto* micro-state without international recognition. It survives with economic assistance from Ankara and a strong Turkish army presence. Fewer than 800 Greeks still reside in the north, mainly around Dipkarpaz (formerly Rizokarpaso) at the tip of the Kirpas peninsula.

The *de facto* partition, or so-called 'Attila Line', extends 180 km from the Kokkina enclave and Kato Pyrgos in the northwest to just south of Famagusta in the east. Between the cease fire lines of each side U.N.F.I.Cyp. controls a Buffer Zone, running the full length of the partition and varying in width from 7 km at its widest point to just 20 m at its narrowest within the city walls of Nicosia, accounting for approximately 3% of the total land area of Cyprus. It cuts across the southern extension of the Morphou citrus groves, slices through the cereal-growing Mesaoria Plain, truncates pre-1974 routeways, and divides important natural water resources.

Central Nicosia has been completely dissected by intercommunal conflict since December 1963, but the events in 1974 meant that the Green Line dividing the city became an even more rigid border. Street blockades guarded by sentry posts were erected on either side of the Line and deserted bullet-scarred buildings, clearly visible along it, are constant reminders of past violence between the communities. There is no civilian Cypriot movement across the Green Line without prior approval from the authorities of both sides and under U.N. supervision, and there are tight restrictions on tourist mobility across it. Occasionally U.N.F.I.Cyp. has arranged family reunions at the Ledra Palace Hotel within the Buffer Zone for those Greeks and Turks residing on the 'wrong' side of the partition, and it continues to pay humanitarian visits to those isolated communities.

Running through the Buffer Zone is the U.N. designated Farming Security Line enabling farmers of either community to utilise land 'between the lines' according to the principles of ownership and security. Although the line is not a permanent ethnic interface, Greeks are not permitted to cultivate land north of it or Turks south of it. U.N.F.I.Cyp. are willing to alter its course in order to extend cultivation provided it is considered safe for both sides to do so. Thus the strip of land separating the two communities is becoming productive again under U.N.F.I.Cyp. auspices.

Meanwhile the political reunification of the island is increasingly unlikely, for the north has undergone many changes since 1974. Its dependence on Turkey for economic and strategic survival was consolidated by the presence of some 20–30,000 soldiers and an influx of 40,000 settlers from the mainland. Turkish is the 'official' language of the north – Nicosia is Lefkoşa; Famagusta is Gazi Magosa; Kyrenia is Girne; and Morphou is Güzelyurt on Turkish maps. On 15 November 1983, Rauf Denktas, the president of the north, proclaimed independence of the Turkish Republic of Northern Cyprus. The renamed state has only been recognised by Turkey. Inter-communal deadlock continues with little progress in discussions about possible federal solutions, or Turkish Cypriot territorial concessions to the Greeks, in spite of U.N.-inspired initiatives to negotiate a settlement.

KEY REFERENCES

Kyle, K., *Cyprus*, Minority Rights Group Report no. 30 (M.R.G., London, 1984).

Patrick, R. A., *Political Geography and the Cyprus Conflict, 1963–1971* (University of Waterloo Press, Ontario, 1976).

Gazetteer

All figures given are for 1983 unless otherwise indicated.

Algeria

Official name	Democratic and Popular Republic of Algeria
Land area	2,381,745 km^2 (919,591 square miles)
Population	20,499,277
Capital and population	Alger (Al Djazair) (1,721,607)
Official language	Arabic
Other language	French
Religion	Sunni Islam
Unit of currency	Algerian dinar
Per capita G.N.P. (U.S.$)	2320

Bahrain

Official name	State of Bahrain
Land area	660 km^2 (255 square miles)
Population	371,000 (1981)
Capital and population	Manama (121,986)
Official language	Arabic
Other language	English
Religion	Shi'ite Islam 60%; Sunni Islam 40%
Unit of currency	Bahrain dinar
Per capita G.N.P. (U.S.$)	10,510

Cyprus (South)

Official name	Republic of Cyprus
Land area	5696 km^2 (2277 square miles)
Population	657,300 (1984)
Capital and population	Nicosia (149,100)
Official language	Greek
Other languages	English, Turkish
Religion	Greek Orthodox Christianity
Unit of currency	Cyprus pounds
Per capita G.N.P. (U.S.$)	3680

Cyprus (North)

Official name	Turkish Republic of Northern Cyprus
Land area	3555 km^2 (1295 square miles)
Population	157,984 (1984)
Capital and population	Lefkoşa (Nicosia) (68,286)
Official language	Turkish
Other languages	English, Greek
Religion	Sunni Islam
Unit of currency	Turkish lira
Per capita G.N.P. (U.S.$)	unknown

Egypt

Official name	Arab Republic of Egypt
Land area	1,000,253 km^2 (386,198 square miles)
Population	44,700,000
Capital and population	Cairo (Al Qahira) (10,000,000 Greater Cairo)
Official language	Arabic
Other languages	English, French
Religion	Sunni Islam 90%
Unit of currency	Egyptian pound
Per capita G.N.P. (U.S.$)	700

Iran

Official name	Islamic Republic of Iran
Land area	1,648,000 km^2 (634,296 square miles)
Population	42,071,000
Capital and population	Tehran (5,734,199)
Official language	Farsi
Other languages	Kurdish, English, French, Azari-Turkish, Arabic, Baluch
Religion	Shi'ite Islam
Unit of currency	Iranian rial
Per capita G.N.P. (U.S.$)	3400

Iraq

Official name	Republic of Iraq
Land area	438,446 km^2 (171,267 square miles)
Population	14,700,000
Capital and population	Baghdad (3,236,000)
Official language	Arabic
Other languages	Turkish, English, Kurdish
Religion	Shi'ite Islam 65%; Sunni Islam 35%
Unit of currency	Iraqi dinar
Per capita G.N.P. (U.S.$)	2148 (1981)

Israel

Official name	State of Israel
Land area	20,700 km^2 (7993 square miles)
Population	4,106,100 plus 1,479,980 in Occupied Territories
Capital and population	Jerusalem (428,668)
Official languages	Hebrew, Arabic
Other languages	English and other European languages
Religion	83% Hebrew; 13% Moslem (exc. Occupied Territories)
Unit of currency	Shekel
Per capita G.N.P. (U.S.$)	5370

Jordan

Official name	Hashemite Kingdom of Jordan
Land area	96,000 km^2 (37,065 square miles) inc. West Bank
Population	2,495,300 (East Bank)
Capital and population	Amman (744,000)
Official language	Arabic
Other language	English
Religion	Sunni Islam – over 80%
Unit of currency	Jordan dinar
Per capita G.N.P. (U.S.$)	1640

Kuwait

Official name	State of Kuwait
Land area	17,818 km^2 (6880 square miles)
Population	1,672,000
Capital and population	Kuwait (775,000)
Official language	Arabic
Other language	English
Religion	Sunni Islam 80%; Shi'ite Islam 15–20%
Unit of currency	Kuwaiti dinar
Per capita G.N.P. (U.S.$)	17,880

Lebanon

Official name	Republic of Lebanon
Land area	10,400 km^2 (4015 square miles)
Population	2,635,000
Capital and population	Beirut (800,000)
Official language	Arabic
Other languages	French, English, Armenian
Religion	Shi'ite Islam 31%; Sunni Islam 21%; Druzes 6%; Maronite Christians 25%; Other Christians 17%
Unit of currency	Lebanese pound
Per capita G.N.P. (U.S.$)	unknown

Libya

Official name	Socialist People's Libyan Arab Jamahiriya
Land area	1,759,540 km^2 (679,358 square miles)
Population	3,356,000
Capital and population	Tripoli (Tarabulus) (481,295)
Official language	Arabic
Other language	—
Religion	Sunni Islam
Unit of currency	Libyan dinar
Per capita G.N.P. (U.S.$)	8480

Morocco

Official name	Kingdom of Morocco
Land area	458,738 km^2 (177,117 square miles)
Population	20,800,000
Capital and population	Rabat (597,000)
Official language	Arabic

Other language	French
Religion	Sunni Islam
Unit of currency	Dirham
Per capita G.N.P. (U.S.$)	760

Oman

Official name	Sultanate of Oman
Land area	271,950 km^2 (105,000 square miles)
Population	1.1–2 m (no official census)
Capital and population	Muscat (85,000)
Official language	Arabic
Other language	English
Religion	Ibadi Islam 75%; Sunni Islam 25%
Unit of currency	Omani riyal
Per capita G.N.P. (U.S.$)	6250

Qatar

Official name	State of Qatar
Land area	11,000 km^2 (4247 square miles)
Population	281,000
Capital and population	Doha (180,000)
Official language	Arabic
Other language	English
Religion	Sunni Islam
Unit of currency	Qatari riyal
Per capita G.N.P. (U.S.$)	21,210

Saudi Arabia

Official name	Kingdom of Saudi Arabia
Land area	2,400,930 km^2 (927,000 square miles)
Population	10,421,000
Capitals and population	Riyadh (Royal) (666,840); Jeddah (Administrative) (561,104)
Official language	Arabic
Other language	English
Religion	Sunni Islam (majority)
Unit of currency	Saudi riyal
Per capita G.N.P. (U.S.$)	12,230

Sudan

Official name	Democratic Republic of Sudan
Land area	2,505,825 km^2 (967,500 square miles)
Population	20,564,364
Capital and population	Khartoum (476,218)
Official language	Arabic
Other language	English
Religion	Sunni Islam, with Christianity and Animism in south
Unit of currency	Sudanese pound
Per capita G.N.P. (U.S.$)	400

Syria

Official name	Syrian Arab Republic
Land area	185,680 km^2 (71,772 square miles)
Population	9,606,000
Capital and population	Damascus (Greater) (2,500,000)
Official language	Arabic
Other languages	French, English, Turkish, Armenian
Religion	Islam – majority Sunni
Unit of currency	Syrian pound
Per capita G.N.P. (U.S.$)	1260

Tunisia

Official name	Republic of Tunisia
Land area	164,150 km^2 (63,362 square miles)
Population	6,800,000
Capital and population	Tunis (596,695)
Official language	Arabic
Other language	French
Religion	Sunni Islam
Unit of currency	Tunisian dinar
Per capita G.N.P. (U.S.$)	1290

Turkey

Official name	Republic of Turkey
Land area	779,452 km^2 (300,947 square miles)
Population	51,600,000 (1985)
Capital and population	Ankara (1,877,755)
Official language	Turkish
Other languages	Kurdish, Arabic, English, French, German
Religion	Sunni Islam
Unit of currency	Turkish lira
Per capita G.N.P. (U.S.$)	1240

U.A.E.

Official name	United Arab Emirates
Land area	92,100 km^2 (32,300 square miles)
Population	1,206,000
Capital and population	Abu Dhabi (296,000)
Official language	Arabic
Other language	English
Religion	Sunni Islam 80%; Shi'ite Islam 20%
Unit of currency	Dirham
Per capita G.N.P. (U.S.$)	22,870

Yemen A.R.

Official name	Yemen Arab Republic
Land area	195,000 km^2 (73,300 square miles)
Population	7,600,000
Capital and population	San'a (277,818)
Official language	Arabic
Other language	English
Religion	Sunni Islam 50%; Sh'ite Islam 50%
Unit of currency	Yemeni riyal
Per capita G.N.P. (U.S.$)	550

P.D.R. Yemen

Official name	Peoples Democratic Republic of Yemen
Land area	336,829 km^2 (130,069 square miles)
Population	2,158,000
Capital and population	Aden (264,326)
Official language	Arabic
Other language	English
Religion	Islam
Unit of currency	South Yemen dinar
Per capita G.N.P. (U.S.$)	520

SOURCES

Middle East and North Africa 1986 (Europa Publications, London, 1986).
Middle East Review (A.B.C. World of Information, Saffron Walden, 1985).
Paxton, J. (ed.), *The Statesman's Yearbook 1984–5* (Macmillan, London, 1984).
World Bank, *World Development Report 1985* (World Bank/Oxford University Press, New York, 1985).

Bibliography

Figures in bold type refer to those maps in the Atlas using that particular source. G indicates general usage.

Atlases

Academy of Sciences, *Physical-Geographic Atlas of the World*. Academy of Sciences of the U.S.S.R. and Main Administration of Geodesy and Cartography of the State Geological Committee of the U.S.S.R., Moscow, 1964. **(3, 4, 8)**

Ady, P. H., *Oxford Regional Economic Atlas: Africa*. Clarendon Press, Oxford, 1965. **(3, 4, 5, 6, 10, 11, 12, 36, 37, 42)**

Air Routes Atlas: May–Nov. 1984. A.B.C. Publications, Dunstable, 1984. **(45)**

Amiran, D. H. K. *et al.* (eds.), *Atlas of Israel*, 2nd edn. Survey of Israel, Jerusalem, 1970. **(55, 13)**

Bartholomew, J., *The Times Atlas of the World*, mid-century edn. The Times Publishing Company, London, 1956. **(G)**

Bartholomew, J. C. (ed.), *The Times Concise Atlas of the World*. Times Books, London, 1985. **(G, 42, 43)**

Bartholomew, J. C. (ed.), *The Times Atlas of the World*. Times Books, London, 1985. **(G)**

Barraclough, G. (ed.), *The Times Atlas of World History*, 1st edn. Times Books, London, 1978. **(14, 15, 16, 17)**

Barraclough, G. (ed.), *The Times Atlas of World History*, rev. edn. Times Books, London, 1984. **(14, 15, 16, 17, 55, 57)**

Bindagji, H. H., *Atlas of Saudi Arabia*. Oxford University Press, Oxford, 1978. **(G, 36)**

Brice, W. C., *An Historical Atlas of Islam*. E. J. Brill, Leiden, 1981. **(14, 15, 16)**

Chaliand, G. and Rageau, J. P., *Atlas Stratégique*. Fayard, Paris, 1983. **(1, 2)**

Committee for the World Atlas of Agriculture, *World Atlas of Agriculture*. Instituto Geographico de Agostini, Novara, Italy, 1969. **(28, 29)**

Couper, A. D. (ed.), *The Times Atlas of the Oceans*. Times Books, London, 1983. **(32, 44, 47, 48, 49, 50, 51)**

Dempsey, M., *The Daily Telegraph Atlas of the Arab World*. The Daily Telegraph, London, 1983. **(G)**

Dixon, C. J., *Atlas of Economic Mineral Deposits*. Chapman and Hall, London, 1979. **(36)**

Dowding, D., *An Atlas of Territorial and Border Disputes*. New English Library, London, 1980. **(52)**

Economist Intelligence Unit, *Oxford Regional Economic Atlas of the Middle East and North Africa*. Oxford University Press, London, 1960. **(5, 6, 9, 10, 11, 12, 28, 29, 36, 37, 42, 43)**

F.A.O. Dept. of Fisheries, *Atlas of the Living Resources of the Seas*. U.N.F.A.O., Rome, 1972. **(32)**

Gilbert, M., *The Arab–Israel Conflict: Its History in Maps*, 3rd edn. Weidenfeld and Nicolson, London, 1979. **(55, 56)**

Griffiths, I. L., *An Atlas of African Affairs*. Methuen, London, 1984. **(37, 38)**

Kinder, H. and Hilgemann, W., *The Penguin Atlas of World History*, vols. 1 and 2. Penguin Books, Harmondsworth, 1974–8. **(14, 15, 16, 17)**

Jones, D. B. (ed.), *Oxford Economic Atlas of the World*, 4th edn. Oxford University Press, London, 1972. **(G)**

Kidron, M. and Smith, D., *The War Atlas*. Pan Books, London, 1983. **(54, 57)**

Loftas, T. (ed.), *The Atlas of the Earth*. Mitchell Beazley and George Philip, London, 1972. **(G)**

Lloyds of London, *Lloyds Maritime Atlas*. Lloyds, London, 1983. **(44)**

Moor, R. I. (ed.), *The Hamlyn Historical Atlas*. Hamlyn, London, 1981. **(14, 15, 16, 17)**

Philip, G. and Swinborne-Sheldrake, T. (eds.), *The Chamber of Commerce Atlas*. George Philip and Sons, London, 1928. **(16)**

Roolvink, R., *Historical Atlas of the Muslim Peoples*. Djambatan, Amsterdam, n.d. **(14, 15)**

Tehran University, *Historical Atlas of Iran*. Tehran University, Tehran, 1971. **(14, 16)**

Van Chi-Bonnardel, R. *et al.*, *Grand Atlas du Continent Africain*. Editions Jeune Afrique, Paris, 1973. **(7, 37, 38)**

Vilnay, Z., *The New Israel Atlas*. Israel Universities Press, Jerusalem, and H. A. Humphrey, London, 1968. **(55, 56)**

Wildon, A. (ed.), *The Observer Atlas of World Affairs*. George Philip, London, 1971. **(G)**

Maps

Defence Survey Maps, Cyprus. 1:250,000. Series K 502, Sheet N1–36–6/7, edn 7-GSGS. M.O.D., London, 1970. **(58)**

Defence Survey Maps, Cyprus. 1:50,000. Series K 717 (various sheets), edn 1–GSGS. M.O.D., London, 1973. **(58)**

Delval, E. (ed.), *The Muslims in the World* (Map). E. J. Brill, Leiden, 1984. **(19)**

The Geographer, *International Boundary Studies* nos. 1, 2, 3, 6, 9, 10, 18, 25, 27–9, 41, 46, 49, 60, 61, 75, 88, 94, 96, 98–100, 103, 104, 106, 111, 121, 163, 164, 167. U.S. Dept of State, Bureau of Intelligence and Research, Washington D.C., 1961–79. **(48, 52)**

The Geographer, *Limits in the Seas* nos. 12, 18, 20, 22, 24, 25, 32 (5th revision), 49, 58, 61, 63, 89, 94. U.S. Dept of State, Bureau of Intelligence and Research, Washington D.C., 1970–85. **(47, 48, 49, 53)**

Hydrographer of the Navy, Admiralty Charts: *The Persian Gulf* (no. 2858, 1959); *Gulf of Aden and S. Part of Red Sea* (no. 6, 1970); *Strait of Gibraltar* (no. 142, 1971); *Jask to Dubayy and Jasireh-ye Qeshm* (no. 2888, 1980). Hydrographic Dept, M.O.D., Taunton, 1959–80. **(47, 48, 49, 51)**

National Geographic, *A Cultural Map of the Middle East*. National Geographic, Washington D.C., 1972 **(18)**

Pennwell Maps, *Middle East Oil and Gas*. Pennwell Publishing Company, Tulsa, Oklahoma, 1983. **(33, 34, 35)**

Survey of Cyprus Administration and Road Map (rev. edn) 1:250,000 Series K 717. Dept of Lands and Surveys, Nicosia, 1981. **(58)**

Suez Canal Authority, *Yearly Report 1984*. S.C.A., Ismailia, 1985. **(50)**

Suez Canal: The Ceremony of the Third Opening. State Information Service, Cairo, 1980. **(50)**

Turkey Highway Map. And Kartpostal ve Yayinlari, Ankara, *c.* 1980. **(43, 49)**

Statistics

A.B.C. Passenger Shipping Guide, March 1985. A.B.C. Travel Guides, Dunstable (monthly). **(44)**

A.B.C. World Airways Guide, Dec. 1985. A.B.C. Travel Guides, Dunstable (monthly). **(45)**

The Bible Societies of the World Annual Report for the Year 1983. United Bible Societies, Stuttgart, 1984. **(19)**

British Petroleum, *Statistical Review of World Energy, 1983–5*. B.P., London (annual). **(46)**

F.A.O., *Fisheries Yearbook 1977*. U.N.F.A.O., Rome, 1977. **(32)**

F.A.O., *Fisheries Statistics 1984* (pre-publication form). U.N.F.A.O., Rome, 1984. **(32)**

F.A.O., *Production Yearbook 1983*. U.N.F.A.O., Rome, 1984. **(29, 30, 31)**

I.L.O., *Yearbook of Labour Statistics*. International Labour Office, Geneva, 1984. **(27)**

Lloyds of London, *Lloyds Register of Shipping*. Lloyds, London, 1984. **(44)**

The Military Balance 1984/5. International Institute for Strategic Studies, London, 1984. (54)

Statistical Abstracts of all states (where available), various years. (G)

Suez Canal Authority, *Suez Canal Report* 1983–5, S.C.A., Ismailia (monthly). (50)

U.N., *United Nations Demographic Yearbook*. United Nations, New York, 1984. (21, 22, 23)

U.N., *United Nations Demographic Yearbook: Historical Supplement*. United Nations, New York, 1979. (21, 22, 23)

U.N., *Statistical Yearbook 1981*, 32nd issue. United Nations, New York, 1983. (G)

U.N., *Statistical Yearbook 1982*, 33rd issue. United Nations, New York. 1985. (G)

U.N., *Yearbook of International Trade Statistics 1982*. United Nations, New York, 1984. (40, 41)

U.N.E.S.C.O., *Statistical Yearbook 1981*. U.N.E.S.C.O., Paris, 1981. (20)

Willet, B. M. (ed.), *The Geographical Digest 1984*. George Philip, London, 1984. (26, 40, 41)

World Bank, *World Development Report*, 1981–5. World Bank/Oxford University Press, New York (annual). (21, 22, 23, 24, 26, 27, 37, 39)

World Tourism Organisation, *World Travel and Tourism Statiscics*, vol. 37, *1982–3* (World Tourism Organisation, Madrid, 1984). (38)

Books and theses

Al-Moosa, A. R. and McLachlan, K., *Immigrant Labour in Kuwait*. Croom Helm, London, 1985. (24, 25)

Amin, S. H., *International and Legal Problems of the Gulf*. Menas Press, London, 1981. (48)

Area Handbook Series (various authors). Country Studies, *Algeria, Cyprus, Egypt, Iran, Iraq, Israel, Jordan, Lebanon, Libya, Morocco, Persian Gulf States, Saudi Arabia, Sudan, Syria, Tunisia, Turkey, The Yemens*. The American University, Washington D.C., 1960–83. (G)

Bacharach, J. L., *A Middle East Studies Handbook*. Cambridge University Press, New York, and University of Washington, Seattle, 1984. (G)

Beaumont, P., Blake, G. H. and Wagstaff, J. M., *The Middle East: A Geographical Study*. Wiley, Chichester, 1976. (G, 7)

Beaumont, P. and McLachlan, K. S. (eds.), *Agricultural Development in the Middle East*. Wiley, Chichester, 1985. (12, 28, 29)

Birks, J. S. *et al.*, 'Who is migrating where? An overview of international labour migration in the Arab world', in A. Richards and P. Martin (eds.), *Migration, Mechanization and Agricultural Labour Markets in Egypt*. Westview Press, Boulder, Colorado and Croom Helm, London, 1983. (24, 25)

Birks, J. S. and Sinclair, C. A., *Arab Manpower: The Crisis of Development*. Croom Helm, London, 1980. (24, 25)

Bowen-Jones, H. *et al.* (eds.), *The Middle East Yearbook*, 1977 and 1978. I.C. Publications, London (annual). (G, 21, 22)

Bowen-Jones, H. and Dutton, R., *Agriculture in the Arabian Peninsula*. E.I.U. Special Report no. 145. Economist Intelligence Unit, London, 1983. (12, 28)

Brant, E. D., *Railways of North Africa*. David and Charles, Newton Abbot, 1971. (42)

Britannica Book of the Year. Encyclopaedia Britannica, Chicago, 1984. (38)

Brownlie, I., *African Boundaries: A Legal and Diplomatic Encyclopaedia*. Hurst and Co., London and University of California Press, Berkeley for Royal Institute of International Affairs, 1979. (52)

Cohen, S. B., *Geography and Politics in a Divided World*. Methuen, New York, 1964. (1, 2)

Drury, M. P., 'The political geography of Cyprus', in H. Bowen-Jones and J. I. Clarke, *Change and Development in the Middle East*. Methuen, London, 1981. (58)

Drysdale, A. D. and Blake, G. H., *The Middle East and North Africa: A Political Geography*. Oxford University Press, New York, 1985. (1, 2, 47, 52)

El-Hakim, A. A., *The Middle Eastern States and the Law of Sea*. Manchester University Press, Manchester, 1979. (53)

Farid, A. M. (ed.), *The Red Sea: Prospects for Stability*. Croom Helm, London, 1984. (53)

Fisher, W. B., *The Middle East*, 7th edn. Methuen, London, 1978. (G, 11)

Gischler, C., *Water Resources in the Arab Middle East and North Africa*. Menas Press, London, 1979. (11, 12)

Glassner, M. I. and de Blij, H., *Systematic Political Geography*, 3rd edn. Wiley, New York, 1980. (1, 2)

Grove, A. T., *Africa*, 3rd edn. Oxford University Press, Oxford, 1978. (10)

Grundy-Warr, C. E. R., 'A geographical study of the United Nations peacekeeping force in Cyprus 1964–84'. Unpublished M.A. thesis, University of Durham, 1984. (58)

Heathcote, R. L., *Arid Lands, Their Use and Abuse*. Longmans, London, 1983. (13)

Karkar, Y. N., *Railway Development in the Ottoman Empire 1856–1914*. Vantage Press, New York, 1972. (42)

Kyle, K., *Cyprus*, Minorities Rights Group Report no. 30. M.R.G., London, 1984. (58)

Luciani, G. (ed.), *The Mediterranean Region*. Croom Helm, London, 1984. (53)

McCaslin, J. C., *International Petroleum Encyclopaedia*, vol. 14. Pennwell Publishing, Tulsa, Oklahoma, 1982. (33, 34, 35)

The Middle East and North Africa. Europa Publications, London, 1983–6 (annual). (G, 19)

Middle East Annual Review. A.B.C. World of Information, Saffron Walden, 1978–80. (G, 11, 42)

Middle East Review. A.B.C. World of Information, Saffron Walden, 1981–5. (G)

Middle East and North Africa: The Strategic Hub. U.S. Army, Washington D.C., 1973. (G)

Mitchell, J. K. M., 'The role of irrigation in the agricultural development of Syria'. Unpublished M.A. thesis, University of Durham, 1982. (11, 12)

Owen, R., *Migrant Workers in the Gulf*, Minority Rights Group Report no. 68. M.R.G., London, 1985. (24, 25)

Patrick, R. A., *Political Geography and the Cyprus Conflict 1963–1971*. University of Waterloo Press, Ontario, 1976. (58)

Paxton, J. (ed.), *The Statesman's Yearbook 1981–2*. Macmillan, London, 1981. (19)

Pitcher, D. F., *An Historical Geography of the Ottoman Empire*. E. J. Brill, Leiden, 1972. (15)

Ports of the World. Benn Publications, London, 1984. (44)

Prescott, J. R. V., *The Maritime Political Boundaries of the World*. Methuen, London, 1985. (53)

Sayin, S. *et al.*, *Atatürk Barajı: Özel Sayisi*. Devlet Su Işleri Bulteni, Ankara, 1983. (11, 12)

Serageldin, I. *et al.*, *Manpower and International Labour Migration in the Middle East and North Africa*. Oxford University Press, Oxford, 1983. (24, 25)

Spykman, N., *The Geography of Peace*. Harcourt, Brace and Co., New York, 1944. (2)

Wagstaff, J. M., *The Evolution of Middle Eastern Landscapes: An Outline to A.D. 1840*. Croom Helm, London, 1985. (14, 15, 42)

The World of Learning 1983–84, 34th edn. Europa Publications, London, 1983. (20)

Key articles and journals

Abella, M. I., 'Labour migration from south and south east Asia: some policy issues', *International Labour Review*, 123 (4), 1984, pp. 491–506. (**24, 25**)

Amin, S. H., 'The Iran–Iraq war: legal implications', *Marine Policy*, July 1982, pp. 193–218. (**48**)

Arab Shipping Guide, 8th edn. Seatrade Publications, Colchester, 1985. (**44**)

Arnold, F. and Shah, N. M., 'Asian labour migration to the Middle East', *International Migration Review*, 18 (2), 1984, pp. 294–318. (**24, 25**)

Blake, G. H., 'Gateway to the Mediterranean', *Geographical Magazine*, 55 (5), 1983, pp. 258–60. (**51**)

Briginshaw, D. *et al.*, 'Railways and railway engineering', business feature, *M.E.E.D.*, 3 Aug. 1984, pp. 25–36. (**42**)

Charnock, A. *et al.*, 'Water resources', business feature, *M.E.E.D.*, 10 Aug. 1984, pp. 27–43. (**7, 11, 12, 28**)

Huan-Ming Ling, L., 'East Asian migration to the Middle East: causes, consequences and considerations', *International Migration Review*, 18 (1), 1983, pp. 19–36. (**24, 25**)

Keesing's Contemporary Archives. Longmans, London, various years. (**G**)

Mackinder, H. J., 'The geographical pivot of history', *Geographical Journal*, 23, 1904, pp. 421–37. (**2**)

Mackinder, H. J., 'The round world and the winning of peace', *Foreign Affairs*, 21 (4), 1943, pp. 595–605. (**2**)

'Middle East oil and gas export routes' (map), *Petroleum Economist*, 51 (7) 1984, pp. 262–3. (**33, 34**)

'The Muslim world', *People*, VI 4 (4), 1979, pp. 1–29. (**19**)

Quarterly Economic Review, annual supplements (all countries of the region). Economist Intelligence Unit, London, 1983–4. (**G**)

Ritchie, M., 'Libya: taking the plunge with the G.M.R.', *M.E.E.D.*, 20, July 1985, pp. 14–16. (**11, 12**)

Serageldin, I. *et al.*, 'Some issues related to labour migration in the Middle East and North Africa', *Middle East Journal*, 38 (4), 1984, pp. 615–42. (**24, 25**)

Many, often anonymous, articles from the following journals were extensively used:

Arab World Agribusiness
Middle East
Middle East Construction
Middle East Economic Digest (*M.E.E.D.*)
Middle East Water and Sewage
World Water